DARK

An A to Z of the Cosmos

JAMES WILKINS

ILLUSTRATIONS BY **ANDREAS BROOKS**

unbound

First published in 2023

Unbound
Level 1, Devonshire House, One Mayfair Place, London W1J 8AJ
www.unbound.com
All rights reserved

Interior art direction by Andreas Brooks and James Wilkins

A CIP record for this book is available from the British Library

ISBN 978-1-80018-229-5 (hardback)
ISBN 978-1-80018-230-1 (ebook)

Printed in China

1 3 5 7 9 8 6 4 2

For Christiana, Xanthe and Polly.

If all there ever was and will be is on this speck of blue,
each night I count my lucky stars to have spent it all with you.

CONTENTS

INTRODUCTION

6

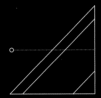

IS FOR
ASTRONOMY

10

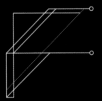

IS FOR
FERMI PARADOX

38

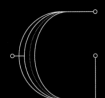

IS FOR
GRAVITY

42

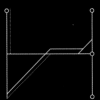

IS FOR
HUBBLE

48

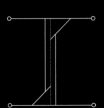

IS FOR
INTERNATIONAL
SPACE STATION

52

IS FOR
NASA

74

IS FOR
OVERVIEW EFFECT

80

IS FOR
PLUTO

84

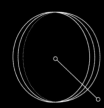

IS FOR
QUANTUM PHYSICS

90

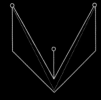

IS FOR
VOYAGER'S GOLDEN
RECORD

112

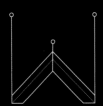

IS FOR
WOMEN AND SPACE

118

IS FOR
X-RISK

122

IS FOR
YELLOW DWARF

126

**IS FOR
BLACK HOLES**

16

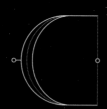

**IS FOR
COPERNICAN
REVOLUTION**

22

**IS FOR
DARK MATTER**

28

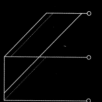

**IS FOR
EXPANSION**

32

**IS FOR
JUPITER**

56

**IS FOR
KUIPER BELT**

62

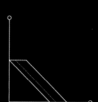

**IS FOR
LARGE HADRON
COLLIDER**

66

**IS FOR
MILKY WAY**

70

**IS FOR
RED PLANET**

94

**IS FOR
SETI INSTITUTE**

98

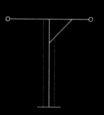

**IS FOR
TRASH**

102

**IS FOR
UNIVERSE**

108

**IS FOR
ZOO HYPOTHESIS**

130

NOTES

136

ACKNOWLEDGEMENTS

150

**A NOTE ON THE
AUTHOR AND
ILLUSTRATOR**

151

INDEX

152

SUPPORTERS

156

Space is kind of a big deal, right? That might be something of an understatement, as space is probably the very biggest deal there is, and guess what? It's getting bigger with every passing second. If that wasn't enough of a mind-melt, our universe could be part of a multiverse, where somewhere out there is a book introduction just like this one, but in Comic Sans type. And another universe, where this intro is very similar but much funnier. Not only is space an extremely big deal, but it's also packed full of wonderful, perplexing, enlightening, thought-provoking and thesaurus-requiring phenomena that will leave you, if not totally speechless, then definitely in need of a broader range of superlative adjectives.

As anyone who is a parent of young children can testify, 'why?' is perhaps the most frequently used word in the household. After looking at the image of Saturn projected on the ceiling from her astro-lamp, my daughter asked me why Saturn wasn't round. I answered that it was fairly round, it just appears a little chubbier around the middle due to its rings. *'Why?'* Because gravity created the rings. *'Why?'* Because gravity is something that... umm... pulls on something else. *'Why?'* Something to do with Newton... and an apple!? Time for bed. And so on and so forth. These conversations highlighted, in no uncertain terms, that I knew far less than I thought I did. My daughter had exposed me to having only a surface-level understanding of the planets, the solar system and the Universe.

And so this project began. Or perhaps I should call it a mission – a mission to increase my knowledge of the Universe even just a fraction so I don't have to google the answer to every question my daughter asks me about space. (Although, in the writing of this book, google furiously I did.) As a layman, I wanted to be able to take my understanding of how the Universe works to a slightly deeper level so in the event anyone – my daughters, my friends, my family – asks 'why'

just a few more times, I might be able to help explain something in a more satisfying fashion. Either that or just be more useful in a pub quiz.

Now let's get something out of the way: I'm not an astrophysicist, cosmologist or astronomer – far from it. However, what I have realised is that space is abundantly accessible for us all. Given a clear evening, one can see the craters of the Moon in extraordinary detail with the most basic of telescopes. Pick the right time and with the naked eye you can watch the International Space Station flying across the sky like a comet. Choose the right location away from the city and the sky will be littered with the light of burning stars in all directions. Whether it's directly observing space from your garden, window or rooftop, or simply settling into your sofa to watch a film that will inspire your thoughts of interstellar travel, wormholes or what life may be like a billion light years away, the Universe is waiting patiently for you to engage with it.

My interest in the Universe has increased exponentially as the years have passed. Naturally, it was the more obviously exciting concepts that piqued my interest initially – black holes, the possibility of alien life, enormous stars that dwarfed our own – but the further I progressed with my own journey of understanding, the more I realised the beauty is as much in the smaller details as it is in the *supermassive.* There are already hundreds of books out there that do a far better job of explaining the cosmos than I will ever be capable of, but this book is trying to achieve something different. I wrote this book as a testament to some of the things I think about when pointing my telescope upward, and to better prepare myself the next time my daughter asks 'why?' It's a modest attempt to tackle a number of space-related points of interest in as succinct a fashion as possible – around 1,000 words a chapter. While it's certainly no replacement for *A Brief History of Time*, this book might help you to think about some of the many awe-inspiring elements that feature in our Universe while reciting your ABCs.

This is *DARK: An A to Z of the Cosmos*. This isn't a science book – it's a *things that caught my interest while googling the word 'space' book*. (Try finding that section in a book shop.) It's an experiment and labour of love to see what happens when I, a non-scientist, attempt to gorge myself on as much space-related information as I could find on the internet, distil the different subjects down to some of their most accessible (and what I found to be the most interesting) essentials and then try and relay that information to others who haven't the time or inclination to read all the articles I did. It's as much about the history, the sociological context and the people around the discoveries as it is about the science itself; a gateway book before you move on to the harder stuff. (A great starting point for said *harder stuff* is to take a look at the sources and references I've included at the end of the book that made my research possible.)

When tasked with choosing only one item of astronomical significance to represent each letter, various problems presented themselves. Some letters are extraordinarily rich in options, meaning tough choices have been made on what to select – B, for example, could have been black holes or the big bang theory; H could have been Higgs boson particle, Hubble, Hadron Collider, Hawking radiation, and so on. I have chosen subjects that represent my personal interest, so I hope that even if you disagree with my selections, you understand that I've done my best to cover as much ground as possible within this construct. And of course, there's no need to read chronologically: you can just dive in and out wherever you choose.

In addition to the visceral mental images that will be conjured from some of the descriptions in this book

(such as how your body would react should it be pulled into a black hole), each chapter is accompanied with an original illustration that represents the subject. The symbiotic relationship between science and art has been known to many of history's great thinkers for centuries; Leonardo Da Vinci proposed that one should 'study the science of art and the art of science'. Many people mistakenly think that you're either a science or an arts person when, in fact, they are inextricably linked, and the more you understand the science, even on a basic level, the more beautiful the concepts become. It was this train of thought that made the visual nature of the book and its various creative expressions as important as the words. Simply put, space is beautiful and therefore, this book should be too.

Since the dawn of our species, we've been looking to the stars for any semblance of meaning. While humanity has produced innumerable feats of genius, furthering our understanding of the cosmos, there is still much left to be understood. While I personally will be little more than a fascinated spectator in this journey of discovery, it is a journey that will continually reveal wonders of unimaginable proportions that we can all benefit from. While most of us will not be individually responsible for scientific discoveries, we can all celebrate the advances and be fully on board for the ride. A great deal of this book was completed during the Covid-19 lockdown, so while my body was largely confined to my house, my mind was taken to the edges of the Universe and to the centre of black holes; it surfed gravitational waves that rolled through millennia and sat atop *Voyager* as it travelled into interstellar space to a Hans Zimmer score.

I hope that this small contribution goes some way to help your mind do the same. While this book is called DARK, it's actually more about light – and to see it, you just have to look up.

'The Universe is full of magical things patiently waiting for our wits to grow sharper.'

EDEN PHILLPOTTS

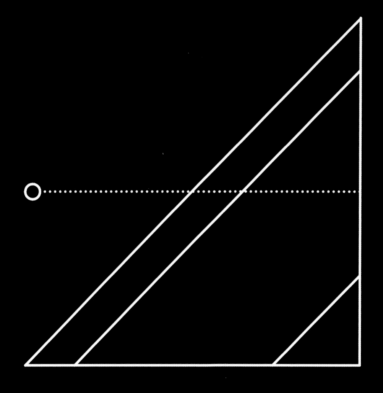

IS FOR

Astronomy

So, let's start at the beginning – the beginning of the alphabet, not the beginning of the Universe. It's rather fortuitous that A stands for astronomy, the very purpose of this book. In short, astronomy is devoted to the study of the Universe and the objects that inhabit it, and is one of the oldest sciences known to humankind. It is both bewildering and awe-inspiring just how adept our ancient ancestors were at unravelling the mysteries of the Universe, with different civilisations drawing their own significant conclusions from the cosmos. For example, research from the universities of Edinburgh and Kent revealed that potentially as far back as 40,000 years ago, cave paintings showed ancient people's knowledge of astronomy through depicting constellations.[1] In the Nubian Desert, a circle of stones thought to have been built more than 7,000 years ago is considered one of the oldest pieces of astronomical apparatus in the world seemingly constructed as an ancient calendar.[2] Sumerian clay tablets revealed that the Babylonians were using sophisticated calculations to map the stars 1,400 years before Europeans, somewhere between 350 BCE to 50 BCE[3] and, while associated with deity worship, the ancient Mayans developed calendars in accordance with the motion of the Sun and stars.[4] The ancient Greeks were pivotal in theorising on the rational design of the Universe, relying on their observations rather than supernatural influences, helping to pave the way for how we see the Universe today. These are just a few examples which illustrate that for as long as humans have walked the Earth, the night sky has been the source of endless speculation, and most civilisations were trying to make sense of it.

Astronomy is sometimes split into two broad remits, observational and theoretical,[5] which work closely together and complement one another. Both branches boast generation after generation of almighty thinkers who have displayed extraordinary methods of observing, predicting and explaining how our Universe might work.

Observational astronomy is concerned with the observation of celestial bodies and matter and the capturing of data relating to the Universe. From the earliest observations using the humble human eye, to modern-day observatories such as the Laser Interferometer Gravitational-Wave Observatory (LIGO), which uses extraordinarily sensitive detectors to witness ripples in space-time, our ability as a species to find ingenious ways to look at the Universe is unrelenting and constantly on the move.

Theoretical astronomy uses mathematics to explain cosmic occurrences and has been frequently used to predict such phenomena prior to any evidence gathered through observational methods. A stunning recent example of theoretical astronomy's predictions would be the aforementioned detection of gravitational waves one hundred years after Einstein theorised their existence in his general theory of relativity in 1916.

While NASA uses the succinct description of astronomy as being 'the study of stars, planets and space',[6] this rather broad term can be expanded upon as it encompasses a space-based smorgasbord of different areas. These include but are by no means limited to the following.

Astrophysics uses physical laws to explain the behaviour of stars and space bodies. According to NASA, the goal of astrophysics is to discover 'how the Universe works, explore how it began and evolved, and search for life on planets around other stars'.[7]

Planetary astronomy focuses specifically on our solar system, covering the planets, planetary systems and other objects that relate to these planets – such as their moons, ring systems and asteroids.

Cosmology deals with the origins of the Universe, which occurred some 13.8 billion years ago (well,

13.77 billion years to be more precise, but what's a few millennia between friends?), and its subsequent evolution. Unless, however, you're a creationist – then it is thought the Universe came to be around 6,000 years ago (but if you're aligned with that, this might not be the right book for you). Cosmology is defined by Britannica.com as the 'field of study that brings together the natural sciences, particularly astronomy and physics, in a joint effort to understand the physical universe as a unified whole'.[8]

Many parents over the decades have encouraged their children to look to the heavens and sing, 'Twinkle, twinkle, little star, how I wonder what you are,' with no reply from the night sky. If a reply were to come, it might be something along the lines of, 'I'm a huge, cataclysmically fiery ball of gas made mainly of hydrogen and helium. Twinkling is a reductive description of my grandeur... and I'm not remotely little.' Stellar astronomy is the study of stars, like our very own Sun, providing insights into the formation of solar systems, the origins of the elements, and how life might form across the Universe.

Astrogeology is the study of the development of planets, their moons, comets, asteroids and other rocky space stuff. Unlike geologists here on Earth, who can get up close and personal with their chosen mineral, astrogeologists must rely on other methods to acquire their data, such as terrestrial telescopes, space telescopes and even rovers such as *Curiosity* and *Perseverance* on Mars.

Are there aliens out there? What might they look like? Are they little, green, or even men?

Astrobiology is the study of how life formed and how life could form in the Universe. Much of this is done on Earth by looking at the development of life in environments with the most extreme conditions. I'm not talking about your average UK kebab shop at 2 a.m. on a Sunday morning. These environments could be within highly acidic volcanic mud pools, by hydrothermal vents on the ocean floor, or embedded in the sub-zero ice of the Antarctic.[9] By looking at the most inhospitable places on the planet, astrobiologists can see what might be necessary to survive in the brutally adverse conditions found on other planets.

Ever heard the phrase 'you really need to broaden your horizons'? Well, that's exactly what practitioners of extragalactic astronomy have done, focusing on the evolution and formation of galaxies outside the Milky Way. The Harvard–Smithsonian Center for Astrophysics reveals that current extragalactic astronomy studies include runaway galaxies, the links between dark matter and supermassive black holes, and how spiral galaxies get their arms.[10]

Astrology is the branch of astronomy that studies large animal-shaped space structures that predict the future, influence our life choices and emotio— Just kidding.

Whether it is using mathematics, a telescope or just pondering the nature of the stars while listening to Pink Floyd, the wonderful thing about astronomy is its accessibility – all you need is a passing interest and, ideally, a clear night sky. So, this is where *DARK: An A to Z of The Cosmos* begins. With a remit encompassing the movements of galaxies, the formation of stars, the behaviour of solar systems and the very origins of the Universe itself, astronomy may well be the biggest, boldest and most important field of study known to humanity.

'The black hole teaches us that space can be crumpled like a piece of paper into an infinitesimal dot, that time can be extinguished like a blown-out flame, and that the laws of physics that we regard as "sacred", as immutable, are anything but.'

JOHN WHEELER

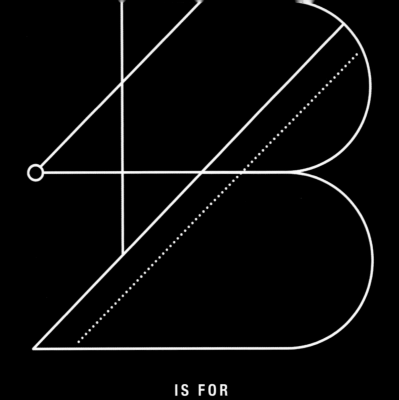

IS FOR

Black Holes

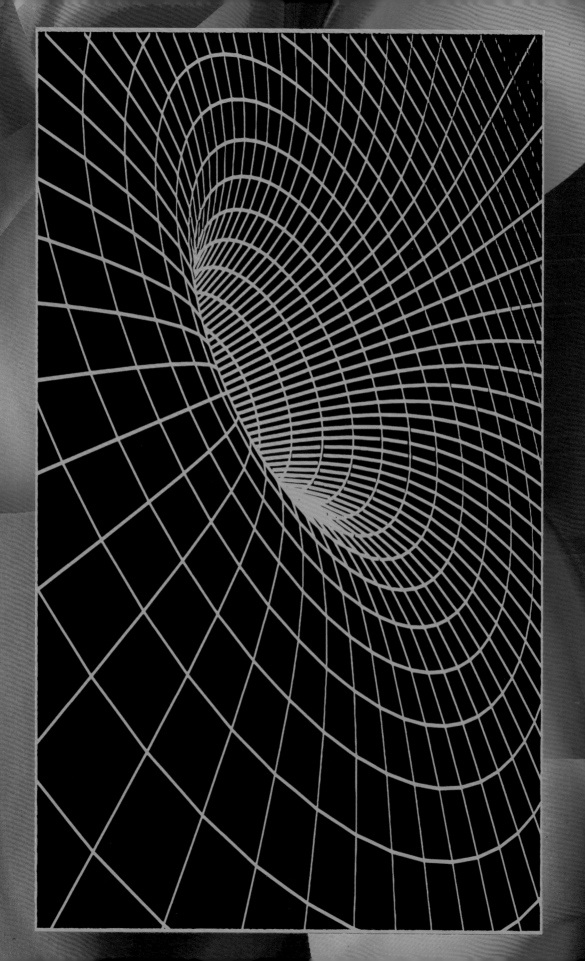

Shadowy monsters of deep space? Star killers? Bottomless pits of doom awaiting hapless planets that stumble in? Black holes could be considered the great white sharks of space. Both are often publicly misrepresented, largely misunderstood and neither is hell-bent on the destruction of life. Not only do black holes pose no imminent danger to us whatsoever, it's highly likely that we owe our very existence to them. Saying that, don't go getting too close to thank them, as they'll still stretch you into nothingness, your malformed body torpedoed into the unknown hell of the singularity, while the rest of the Universe ages in relative high speed before your soon-to-be-dematerialised eyes.

And breathe.

Black holes were first proposed in 1783 by clergyman and philosopher John Michell[1] who, through an insightful thought experiment, theorised that there may be something with enough mass to prevent light from getting away – the extreme forces of gravity too strong even for light to resist. If being a philosophising holy man isn't enough of a mic-drop, being the first person to think up black holes certainly is.

Since then, a number of great minds have developed our knowledge of the mechanics of black holes, including the heavily grey-mattered Albert Einstein with his theory of general relativity. Contrary to Newton, Einstein concluded that gravity wasn't a force that *pulled* on other things. Instead, gravity was a warping of space-time – the greater the mass of 'something' in space, the greater that 'something' will warp space.[2] Once, at my daughter's bedtime, I tried to explain this warping of space-time as the indentation of her mattress from me sitting on it. The greater the space-bottom, the greater the dent

in the space-time mattress. Not entirely accurate, but it helped. Mattresses aside, it's safe to say that the space-time warping caused by black holes changes the very foundations of the known Universe. Like tasting Marmite Peanut Butter for the first time. Most recently, Professor Roger Penrose won half of the 2020 Nobel Prize for Physics for devising an argument verifying Einstein's general theory of relativity in relation to black holes. The University of Oxford's official website commented, 'Roger Penrose used ingenious mathematical methods in his proof that black holes are a direct consequence of Albert Einstein's general theory of relativity. Einstein did not himself believe that black holes really exist.'[3] The other half of the Nobel Prize went to Professor Andrea Ghez and Professor Reinhard Genzel, whose pioneering observational methodologies proved that a supermassive black hole was at the centre of our own galaxy, the Milky Way. If you want to doubt the value of your own life achievements, take it from me, just write a book about space.

A black hole is an area of space where matter has collapsed in on itself, resulting in a truly enormous amount of mass being concentrated in a small area. Like repacking your luggage on the return leg of a holiday. Most black holes are formed when a giant star – many, many times larger than our Sun – starts building up iron at its core, making it unstable. A star's equilibrium is based upon the balance between radiation pushing outwards and gravity holding it together, which maintains the stability. Iron is the proverbial cat among the pigeons, destroying this harmony. In other words, the star kicks the bucket and dies. Dramatically. While our own Sun's death will be a relatively graceful demise, with its outer layers drifting into the cosmos, a far bigger star's death (as a result of iron at its core reaching a critical amount) leads to cosmic violence of cataclysmic proportions –

the outer layers exploding in a supernova, providing some of the brightest activity in the Universe. The very definition of going out with a bang. At the same time, the star's core collapses into itself, forming a gravitational singularity – a place where the known laws of the Universe cease to operate.[4] The very fabric of space-time is warped so severely that not even light can escape and time itself is desecrated.

The imprisonment of light means black holes are invisible to direct observation, which is what led to their name, coined by physicist John Wheeler. Wheeler said in his autobiography that a black hole 'teaches us that space can be crumpled like a piece of paper into an infinitesimal dot, that time can be extinguished like a blown-out flame, and that the laws of physics that we regard as "sacred", as immutable, are anything but'.[5]

If the entrapment of light isn't enough to visualise the almighty gravitational forces that black holes exert, consider *spaghettification.* Which, before you ask, isn't a Dalston eatery. Due to the gravitational forces from a black hole, if one were to enter it feet first, the gravity would pull more powerfully on one's feet than one's head, stretching the human form into something that resembles, yes, you've guessed it, a piece of spaghetti. While the idea of being transformed into a human super-noodle is scary enough, due to gravity's influence on space-time, time (as you experience it) would be running far more slowly the closer you got to the black hole than for everyone else. You'd see the rest of the Universe age before your eyes. Being stretched to death while everything you've ever known hurtles through millennia in a heartbeat does not sound like a pleasant way to go. It'd seem quick to you, but to everyone else, your transformation into a cheese string could take millions of years. How embarrassing.

Having the power to feast on passing stars and keep light as a house guest would imply that a black hole's grip is somewhat irrefutable. However, Stephen Hawking's pioneering research taught us that black holes don't keep everything to themselves: they may actually decay and shrink through the radiation they give off – a process known as Hawking radiation.[6] But the demise of a black hole will not be a quick process. Astronomy and physics professor at Yale University Priyamvada Natarajan said of the decaying black-hole process, 'The entire age of the Universe [is] a fraction of [the time] it would take ... As far as we're concerned, it is eternity.'[7]

Despite their infamous reputation, black holes are integral to the structure of galaxies. It's commonly believed that there's a supermassive black hole at the centre of every galaxy, a shadowy ringleader helping to create stars from the elements thrust into space from their supernova explosions.[8] *The Atlantic*'s Marina Koren pointed out that the riotous explosions that precede a black hole's formation 'spew elements such as carbon, nitrogen, and oxygen into space. The collisions of black holes and neutron stars help spread heavier elements, such as gold and platinum. These elements make up our Earth, and our own selves.'[9] So, if someone tells you they don't like space, they clearly have some self-loathing issues going on.

Despite their being invisible to direct observation, there are ways to *look* at black holes, primarily through the influence that they have on matter in their surrounding neighbourhood – kind of like a passive-aggressive work colleague who isn't openly rude, but you can sense the increasing frustration in those around them. One method is to observe the way light from behind a black hole bends around it, known as 'gravitational lensing'. It's also possible to measure other stars orbiting a black hole to determine its size and position. In an astounding feat of technical wizardry, 2015 yielded further proof of black holes by detecting the ripples in space-time caused by the collision of two black holes 1.3 billion light years away.[10] These cataclysmic collisions disturbed space-time to such an extent that they sent ripples across the Universe which scientists at the Laser Interferometer Gravitational-Wave Observatory were able to measure. These ripples, or gravitational waves, were proposed by Einstein in 1916, who theorised that they would occur as a result of the collision of two massive accelerating objects.

In what may be the most important and impressive photograph of all time, 2017 yielded the first-ever image captured of the Universe's most infamous and intriguing occurrence: a supermassive black hole. The giant in question is 6.5 billion times the mass of our Sun and is residing at the centre of the M87 galaxy, 55 million light years away. Named after the boundary around a black hole from which light cannot escape, the Event Horizon Telescope used to capture the image was, in fact, a network of eight telescopes from across the planet in Hawaii, France, Arizona, Greenland, Chile, Mexico, the South Pole and Spain. Each of the telescopes observed M87 over a seven-day period, resulting in 5 million gigabytes of data[11] that took months to process. But the wait was more than worthwhile, as the world was provided with a breathtaking image of the giant black hole – the very first glimpse of a primordial space-time abyss. However, what lies within a black hole and how the laws of physics work past the event horizon is still unknown.

Nobel Prize-winner Professor Ghez said, 'Our observations are consistent with Einstein's general theory of relativity. However, his theory is definitely showing vulnerability. It cannot fully explain gravity inside a black hole, and at some point we will need to move beyond Einstein's theory to a more comprehensive theory of gravity that explains what a black hole is.'[12] For those of you who, like me, are only just getting your heads around the broad sweeps of Einstein's theories, the prospect that there are many other, as-yet-undiscovered laws governing the Universe may seem daunting. But fear not! Our capacity for understanding can continue to expand, just like the Universe.

Putting the evolving theories regarding gravity aside, black holes are perhaps the most fascinating phenomena in the Universe. A place where space and time cease to make sense. A place where the perceived laws of the Universe are not obeyed. They serve as a reminder of how much there is still to be discovered and understood behind the threshold of the event horizon.

'If you feel you are in a black hole, don't give up. There's a way out.'

STEPHEN HAWKING

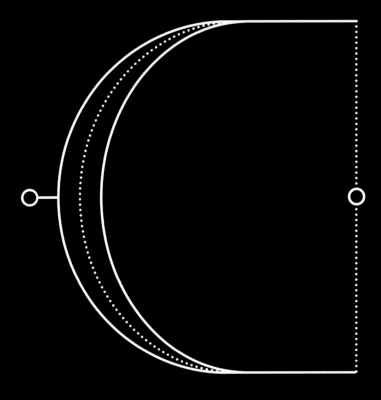

IS FOR

Copernican Revolution

Many moons ago, when taking in the views of the night sky, our ancestors could be forgiven for assuming that the bright objects that appeared to pass above them were circling the seemingly stationary chunk of rock that they were standing on. However, not only were the stars *not circling* our ancestors, but the planet upon which they stood was itself hurtling through space at 107,826 km/h on its very own orbit around the Sun. Plot twist!

The theory that the Earth was at the centre of the Universe is called the geocentric model and it proposed that the Moon, the stars and the Sun all orbited our planet. This model, also known as the Ptolemaic model after the Greek astronomer and mathematician Claudius Ptolemy, would have seemed a perfectly reasonable assumption at the time. After all, even with all the knowledge we possess in modern times, we cannot 'feel' any movement of our planet, let alone 107,800 kph of movement. To Ptolemy, Earth seemed perfectly still and the Sun, the Moon and the stars all appeared to circle it once a day. Geocentrism, therefore, was the dominant theory in ancient Greece, and it wasn't until 1543 that Nicolaus Copernicus proposed a rather different idea, one that would dramatically change our collective point of view. In his book *On the Revolutions of Heavenly Spheres*,[1] Copernicus argued that it was in fact the Sun that was at the centre of the solar system, with the Earth and the other planets orbiting around it. But Copernicus wasn't the first to theorise a heliocentric model. His work followed a number of other scientists, including two prominent Islamic astronomers – Mu'ayyad al-Din al-Urdi in the thirteenth century and Ibn al-Shatir in the fourteenth century[2] – who also disagreed with Ptolemy's model and sought to rectify the inconsistencies. In fact, the first known suggestion that the Sun may be at the centre of the known Universe was made by ancient Greek astronomer and mathematician Aristarchus of Samos, who was born in 310 BCE.[3]

Despite the theory of a heliocentric universe being proposed many hundreds of years prior to 1543, it was Copernicus's theory that brought about a revolution in the dominant scientific and cultural view of our place in the cosmos. Copernicus would very likely have been aware of the controversial nature of his work but perhaps unaware of just how big an impact it would have on astronomy, society, culture and religion. But this revolution did not take place overnight. Over the next century, increasing evidence from the scientific community built upon Copernicus's theory became much harder to deny. Let's take a moment to talk about three of the most important scientists to help perpetuate this shift in worldview: Galileo Galilei, Johannes Kepler and Sir Isaac Newton.

Possessing opinions that went against the Church's assertion that we, as God's creation, were at the centre of the Universe, were often met with dire, if not fatal, consequences. To give some context on the potential ramifications for heresy at the time, 1609–14 gave rise to the Basque Witch Trials, which arose from the largest witch hunt in history and saw as many as 7,000 people accused of witchcraft. According to *Smithsonian Magazine*, 'at least 2,000 of those accused were investigated and tortured, and 11 died. Six were burned at the stake and five were tortured to death in prison.'[4] Drawing attention to yourself for perceived heretical beliefs was a risky business. Around the same time, not too far away, Galileo Galilei was conducting experiments that would result in his own witch hunt. After hearing of the invention of the telescope, Galileo had embarked on creating his own, which he chose to point upwards. It was a choice that resulted in a series of incredible discoveries that compounded his support for Copernican theory. Among Galileo's discoveries were the different phases of Venus; identifying sunspots, which confirmed that the Sun rotates; observing four of Jupiter's moons; observing the craters of our own Moon; discovering many new stars and confirming that the planets rotate around the Sun.[5] Unfortunately, Galileo wasn't rewarded for his efforts: he was accused of heresy, as suggesting that anything other than Earth being at the centre of the Universe went against religious status quo. Galileo was forced to retract his findings and sentenced to life imprisonment. Perhaps it is some small mercy that due to being nearly seventy years old, he was permitted to serve his sentence under house arrest.[6]

German astronomer and mathematician Johannes Kepler was best known for his laws of planetary motion in works including *Epitome of Copernican Astronomy*.[7] These laws furthered Copernican theory, explaining how planets' orbits not only vary in speed but also follow elliptical paths rather than circular. It's important to add that even with a scientific mind as potent as Kepler's, the influence of religion was equally powerful. A deeply religious man, Kepler was always trying to find the balance between science and his beliefs. His primary job was as an astrologist, so he strove to justify heliocentrism on religious grounds as much as through scientific analysis.[8] In Kepler's case, the battle between science and religion had a very personal resonance, as his own mother had been accused of witchcraft. Due to numerous spurious accusations levelled at Katharina

Kepler, including sightings of her turning into a cat, she was arrested and spent six years incarcerated. Kepler sidelined his own work and moved his family to help defend his mother. His legal case, described in an article from the University of Cambridge as 'a rhetorical masterpiece',[9] resulted in his mother's eventual exoneration. Unfortunately, she died just six months later.[10]

Building upon the works of Copernicus, Kepler and Galileo, the publishing of Newton's *Philosophiae Naturalis Principia Mathematica (Mathematical Principles of Natural Philosophy)*[11] in 1687 could be perceived as the pinnacle of the world-view shift from geo- to heliocentrism. Newton's laws of motion and universal gravitation obliterated any final doubts surrounding the heliocentric model, putting forward principles that would account for the movements of both Earth and other celestial bodies.

It's difficult to overstate just how much of an outrageous challenge to the social norms it would have been to suggest that we, on Earth, were not at the centre of the Universe. It went against everything people had been taught, changed how we saw ourselves and undermined the religious status quo. This revolution may not have been sudden, explosive, or attributable to a single moment, but it truly reflected humankind's desire to ask questions, however inconvenient and disruptive they may be.

Q

Why did Ptolemy's wife leave him?

A

Because he thought the Universe revolved around him.

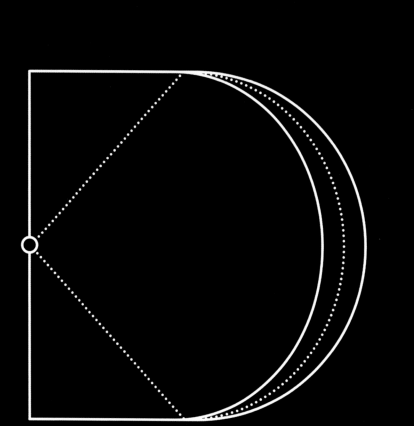

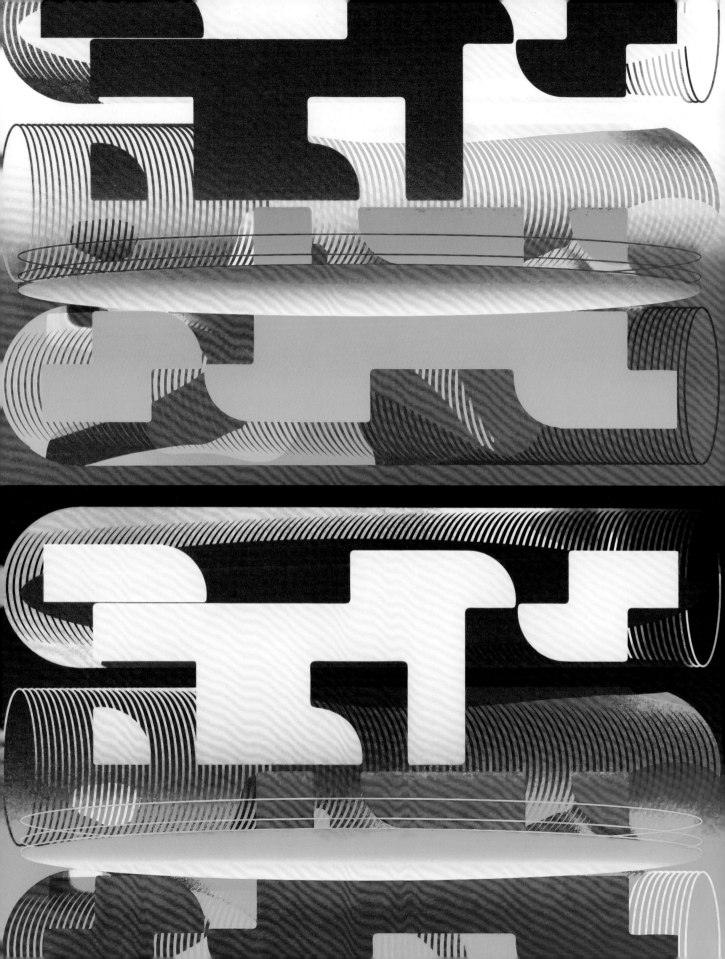

When you look up on a clear, cloudless evening, the sky can appear positively ablaze with the light from billions of stars shining down upon us. However, it's strange to think that there's something else up there that accounts for far more of the Universe's matter than all the stars and galaxies combined. This chapter is about something that cannot be seen, cannot be directly measured and while we know what it *does*, we don't know what it *is*. 'What on Earth is the matter?' is an exasperated question inhabiting relationships worldwide – but when it comes to the Universe, it takes on a whole different meaning.

Dark matter accounts for around 27 per cent of the known mass of the Universe but it has never been directly observed. In fact, all the galaxies, stars and planets make up just 4.9 per cent of the Universe's mass energy. So, if all the observable galaxies, stars and planets make up 4.9 per cent, and dark matter accounts for around 27 per cent, then what constitutes the rest of the Universe?[1] We'll come on to that in 'E for Expansion' and a little ol' thing called dark energy.

The ominously titled dark matter was given its name due to it not giving off, absorbing or reflecting light, rendering it invisible to the electromagnetic spectrum. However, despite this cloak of invisibility, there are some tell-tale signs that dark matter exists.

Similarly to black holes, evidence for dark matter includes its gravitational effects on visible matter, such as the movements and structure of the galaxies and also gravitational lensing, where light is bent as it passes through the gravitational warping of space-time. Due to the fact that things that are dark absorb light, theoretical physicist Lisa Randall considers 'transparent matter' to be a more accurate description.[2] Due to its reluctance to reveal itself while also showing its gravitational influence, I would like to propose 'coy matter' as a suitable alternative to help give this mysterious matter a much-needed PR overhaul. (There is probably a very good reason why nobody has asked me to officiate over the naming of important astrophysical anomalies.)

The first suggestion of dark matter came about in the 1930s through the separate observations of Jan Oort and Fritz Zwicky. In 1932 Jan Oort observed that stars in our galactic locality were moving too fast and thought that there must be more mass than was visible holding the galaxy together. He suggested that the missing mass might be due to darker stars that emitted too little light to be observed.[3] A year later, Zwicky observed a cluster of around 1,000 galaxies called the Coma Cluster and made an estimation of the amount of mass contained within it. He also measured the velocities of some of the galaxies and, similarly to Oort, saw that they were moving faster than the total mass of the observable matter should be able to support.[4]

It wasn't until the pioneering work of Vera Rubin in the 1960s and 1970s that the case for dark matter was taken seriously. Rubin's studies of spiral galaxies found their outer edges to be travelling faster than expected – as fast as their centres. According to the observable mass within the galaxies, this quite simply shouldn't be possible. Newtonian physics dictates that the stars on the outside of a galaxy should be moving more slowly than those at the centre, which are feeling the effects of gravity more intensely.[5] However, stars appear to move at roughly the same speed throughout galaxies, which means there must be way more mass in a galaxy than we can actually see, acting as a cosmic glue and holding them together. If the galaxies were only bound by the gravitational effects from their own observable mass (like planets and stars), they would have been pulled apart a long, long time ago. Rubin used the dark-matter conundrum as a metaphor for how well we understand the cosmos as a whole, commenting, 'In a spiral galaxy, the ratio of dark-to-light matter is about a factor of ten. That's probably a good number for the ratio of our ignorance-to-knowledge. We're out of kindergarten, but only in about third grade.'[6]

Despite our knowledge of how important dark matter is to the composition of galaxies, it is still seemingly shrouded in cosmic camouflage. Common theories suggest that it may be WIMPs (Weak Interacting Massive Particles, hypothetically created early in the lifespan of the Universe) that interact with other forms of mass but neither give off nor absorb light.[7] If this is the case, then it's just a matter of time before detectors of sufficient sensitivity are developed to measure such incredibly puny interactions. Alternatively, it could well be that the current understanding of gravity is incorrect and what we currently assume is dark matter is a product of this misunderstanding. Other emerging theories refer to 'supersymmetry and extra dimensions'[8] – concepts that stray outside of the current standard model of physics and would require an author of considerably higher IQ to recount.

Bill Bryson succinctly commented in his book *A Short History of Nearly Everything*, 'For the moment we might very well call them DUNNOS (for Dark Unknown Nonreflective Nondetectable Objects Somewhere.'[9] For now, the prospect of directly observing dark matter or understanding its precise nature remains out of our grasp. But what we do know is that whether it's an as-yet-undiscovered particle, or even an indication that our current understanding of gravity is incorrect, the unravelling of dark matter's secrets will continue to challenge scientists for many years to come, and is another addition to the humble list of the Universe's great unknowns.

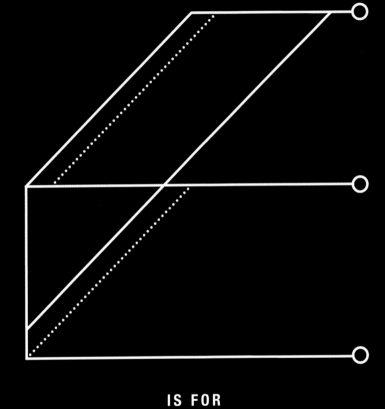

IS FOR

Expansion

The fact that the Universe is constantly expanding in all directions is one of the most disconcerting, disorientating and discombobulating discoveries in astronomy, reinforcing the knowledge of our somewhat insignificant place in the Universe. It has not only inspired scientists to speculate on why it is expanding and where this may lead, but it has also allowed us to rewind the journey of space-time and see the very origins of the Universe.

So, if dark matter accounts for around 27 per cent of the Universe and observable mass accounts for a measly 4.9 per cent, that leaves another 68 per cent unaccounted for. Missing in action. AWOL. How can something so ubiquitous be so taciturn? This is where dark energy reveals itself, or, if we're being pedantic, doesn't reveal itself. Not only does dark energy seemingly exist evenly throughout the Universe but it is also theorised to be causing the expansion of the Universe. For those of you with a fear of high speeds, roller coasters or even mildly aggressive merry-go-rounds, try not to think about the fact that this expansion also appears to be accelerating. All the time. For ever.

In the early 1900s, one astronomer's observations of spiral galaxies provided the first evidence supporting the idea that the Universe was expanding.[1] The astronomer, named Vesto Slipher, reported that nearly all the spirals he observed were moving away from us, something he deduced from the redshifting of the light.[2] Redshift is a feature of the Doppler effect, and happens when the wavelength of light emitted by an object moving away from the observer is stretched, and is shifted towards the red end of the spectrum. If we can see that the light from distant galaxies is shifted in this way, then we know it is moving away from Earth.[3] In 1929 Edwin Hubble announced that the further away the galaxies were, the faster they were moving. Hubble placed a specific method against measuring the rate of this expansion, which is now called Hubble's law. Rather than all the galaxies in the Universe moving further apart *through* space, it is the fabric of space itself that is expanding, increasing the distance between galaxies. Visualising things that are incomprehensibly big and seemingly expanding into an unfathomable void is tricky, but astronomers often use the example of a balloon to help visualise the Universe's expansion. Imagine dots on the balloon's surface and as the balloon is inflated, the dots are stretched further and further apart – the Universe, it is thought, is not expanding into new areas, merely expanding in and of itself.[4] But, as with many discoveries of this magnitude, credit is not always clear-cut. Just a few years before Hubble's 1929 announcement, a Belgian priest called Georges Lemaître proposed a solution to Einstein's general relativity equations saying that the Universe should be expanding. This led to the International Astronomical Union voting to amend the name Hubble's law to the Hubble-Lemaître law.[5]

The High-Z Supernova Search Team and the Supernova Cosmology Project might sound like flamboyant 1990s World Wrestling Federation tag-teams, but they were, in fact, the scientific collectives which both found results suggesting that the Universe was expanding at an accelerating rate. A discovery that led to three people from these groups being awarded Nobel Prizes. At the time of awarding the Nobel Prize, the official press release theorised on the future of the Universe: 'What will be the final destiny of the Universe? Probably it will end in ice, if we are to believe this year's Nobel Laureates in Physics. They have studied several dozen exploding stars, called supernovae, and discovered that the Universe is expanding at an ever-accelerating rate. The discovery came as a complete surprise even to the Laureates themselves.'[6]

Prior to this discovery that the Universe was expanding at an accelerating rate, it was expected by scientists that the expansion would be slowing down, due to the effects of gravity. However, upon knowing that this wasn't the case, the scientific community began speculating on what could be causing this accelerated expansion. This cause is commonly thought to be dark energy – but this doesn't actually explain what dark energy is. And truth be told, no one knows. When referring to dark matter and dark energy, astrophysicist Neil deGrasse Tyson said that they may as well be called 'Bert and Ernie', as the term dark energy is merely a temporary name for something that we really know very little about, acting as, 'placeholders for our abject ignorance'.[7]

However, that old adage 'the snapping jaws of discovery eat placeholders for abject ignorance for breakfast' (you've all heard of that saying, right?) couldn't be truer, as there are, in fact, a few theories as to what dark energy is. These include a 'cosmological constant' theory, and another called 'quintessence'. To backtrack somewhat, Einstein may have been the first to suggest an expanding Universe. When trying to apply his theory of relativity to the structure of space and time, he came to the realisation that it didn't work for a static universe. It wasn't known in 1917 that the Universe was expanding, so this led to Einstein introducing a repulsive force, which he called the cosmological constant, to reconcile the theory of relativity with the static universe. When it was revealed that the Universe was expanding, Einstein then removed the cosmological constant from his equations.[8] Writing for *Forbes*, Ethan Siegel said, 'Einstein realized that his theory predicted that a static Universe was unstable, and that it must expand or contract. Rather than accept this robust prediction, though, Einstein instead rejected it, assuming the Universe must be static. Instead, he

introduced his cosmological constant to compensate, leading to what he later referred to as his "greatest blunder" in all of physics.'[9]

Current cosmological constant theory represents a different interpretation of Einstein's original idea. It proposes that there's a constant vacuum energy density that is spread evenly throughout space as a result of particles that continually snap, crackle and pop in and out of previously empty space and fuel the expansion (known as quantum fluctuations).[10] This is in contrast to Einstein's original theory that saw the cosmological constant as an agent of balance, offsetting any expansion.[11]

An alternative proposed theory for the expansion of the Universe is known as quintessence, which is an unknown energy field that overpowers gravity and pushes particles away from one another and, unlike the cosmological constant, can vary and evolve over time.[12]

The problem with theories surrounding the expansion of the Universe is the difficulty with measuring gravity on such an enormous cosmological scale, unlike the way we're able to with solar systems and even galaxies. However, the incredible advancements in our experimental capabilities with facilities like the Large Hadron Collider will increase our chances of unravelling the secrets of the Universe's expansion and, hopefully, much more. Our ability to further our understanding of particle physics will perhaps bring us closer to uncovering a grand unifying theory uniting the laws that govern the very large and the very small, the reconciliation of the cosmic and the quantum. Big and small. Bert *and* Ernie.

'The Universe is a pretty big place. If it's just us, seems like an awful waste of space.'

CARL SAGAN

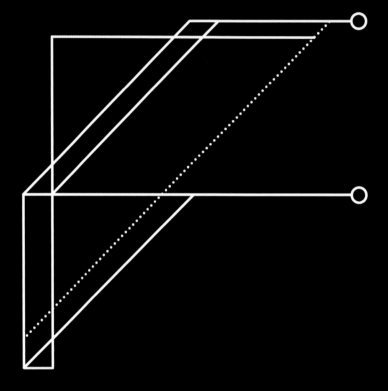

IS FOR

Fermi Paradox

When talking about space, it's easy to get carried away with numbers so large that they soon become entirely meaningless. So I'll try not to dwell for too long on any number with more than, let's say, sixteen zeros in it. However, with at least 100 billion stars in the Milky Way[1] (and that's a conservative estimate) and at least 100 billion galaxies in the known Universe (which will probably increase as observational technology advances), it's mathematically safe to say that Earth should not be the only place to boast life.

Following the Kepler space mission, whose objective was to determine the percentage of planets that are in or near the habitable zone of stars,[2] astronomers suggest that there are as many as 40 billion Earth-sized worlds in the Milky Way alone that orbit within this habitable zone. This 'Goldilocks zone' describes the ideal environment for life, where water can pool on a planet's surface. So, with billions of potentially Earth-like rocky planets in the Milky Way and many, many more in the wider Universe, one Italian physicist, Enrico Fermi, asked the not unreasonable question: where is everybody?[3] The Fermi paradox addresses the conflict between the vast opportunities for life to develop throughout the trillions of Earth-like planets across the Universe and the complete absence of evidence for any life other than that here on Earth.

The Drake equation, created in 1961 by Frank Drake (who sadly died in 2022 at the age of ninety-two), was a mathematical equation that assessed the probability of extraterrestrial civilisations. This was based upon various factors, including the rate at which stars are formed, the number of these stars with planetary systems, the number of planets within these solar systems that possess environments suitable for the development of life, and the fraction of such planets where intelligent life can form and then develop communication technology.[4] Drake's original calculations led him to believe there should be between 1,000 and 100,000,000 planets and civilisations in the Milky Way. As you

might agree, this is quite broad. It'd be like asking someone what time you're meeting for dinner and them replying, 'Hmm, I'll meet you between January 1982 and November 2025.' According to astrophysicist Paul Sutter, 'The Drake equation is simply a way of chopping up our ignorance, stuffing it into a mathematical meat grinder and making a sausage-guess.'[5]

Of course, much has changed over the last fifty years. Our ability to map the stars within the Milky Way has improved dramatically with space telescopes like Kepler modifying the variables in the Drake equation. *National Geographic* reported, 'They found that between 37 and 60 percent of sunlike stars in the Milky Way should host a temperate, Earth-size world — and using a more liberal calculation of the energy needed for a world to be temperate, they found that as many as 58 to 88 percent of sunlike stars could have such a world.'[6] However, there are a great deal of stars far older than our own, meaning the evolution of a technologically advanced species able to traverse our galaxy and make contact should already have happened.

So, are we really screaming into the endless empty void or are there valid reasons for the radio silence and intergalactic ghosting? Well, considering the probability that life (in some form) other than our own should exist in the Universe, there are many explanations to suggest why we have not yet been contacted. I've focused on some of the most compelling arguments below.

The Great Filter theory suggests that some sort of barrier exists that no species can pass, and this metaphorical barrier could take many different forms. For example, creating the right environment for life to form appears to be a lengthy and difficult process. Then, once you have life, there are potentially catastrophic dangers posed by asteroids and comets... as proven by the extinction of the dinosaurs. There are also the dangers species pose to themselves as they become more advanced, be it global warming, the misuse of nuclear power by a paranoid narcissistic despot or emerging technology like dancing robot dogs gone bad. If the great filter does exist, it is unknown whether this barrier exists in our past (in which case we have surpassed it – great news!), or in our future (in which case it awaits us – not such great news). It's important to remember that space is a very difficult place to survive over long periods of time. Asteroid impacts, gamma-ray bursts, solar flares and supernovas[7] are just some of the ways the Universe may unwittingly seek to bring any fledgling civilisation to an end. A smorgasbord of circumstantial sadism. Taking into account the huge timescales we are dealing with, something hideously catastrophic is extremely likely to happen to any given lifeform at some point. Any species that evolved to a point of interstellar communication may have become extinct, either through natural disaster or self-destruction many millions of years before any other species came to be. With a Universe as old as this, it would be easy for two civilisations to miss each other by millions, if not billions of years. If you're an old romantic who believes in fate, this is also presents serious problems for your love life. If you're wating for 'the one,' imagine the possibility that 'the one' won't evolve for another 750 million years on a planet several parsecs away.

As we've discussed, space is enormous – so enormous it's possible that communications from aliens are travelling through the cosmos, but they haven't had enough time to reach us yet. Michael Garrett, director of the Jodrell Bank Centre for Astrophysics, said, 'On average, you'd expect the civilizations to be separated by at least 1,000 light years in the Milky Way. That's a large distance, and for communication purposes you need to allow for twice the travel distance, so you're talking about civilizations that have to be around for at least a few thousand years in order to have the opportunity to talk to each other.'[8] Even if they were to reach us within the very brief time that we've been looking, there's nothing to say that it'll be in a format that we are as yet capable of interpreting as alien communication.

An alternative argument suggests that there may be other species in the cosmos far more intelligent than us, but this chasm in intelligence renders us completely uninteresting and insignificant to them. This is entirely unsurprising, given our species cohabits this planet with thousands of different species but in the most part shows a complete disinterest in any meaningful dialogue with any of them, let alone perceiving any of these other species as equals. But much like zoologists here on Earth, we may well be of interest to alien civilisations from an observational perspective. We'll go into this in a little more detail in 'Z is for the Zoo Hypothesis'.

There are many things to be proud of as a species, but it wouldn't be a huge leap of judgement to assume that we are some way from perfect in the arc of enlightened evolution. Despite the exponential increases in our acquisition of knowledge, we are still a species capable of great damage to ourselves and to our planet, and it often feels like we are a long way from viewing ourselves as a unified species rather than a divided one. With this in mind, the theory that we might be the most advanced civilisation in the Universe is rather terrifying.

Whether we're the most advanced civilisation in the entire Universe or merely stumbling through our species' infancy while far superior beings traverse the galaxies uninterested in our brutish mewling, ostensibly, we are alone. One thing is apparent: despite our rapid and ever-increasing understanding of the Universe and development of sophisticated tools, it's likely that we lack the technological capacity to either project or detect communications with sufficient success to result in meaningful communication any time soon. However, our ongoing endeavours to find life in the cosmos and to advance in technology sufficiently to aid the search will undoubtedly improve our chances... even if the odds are stacked against us.

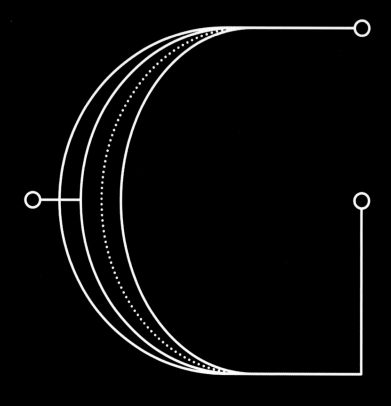

IS FOR

Gravity

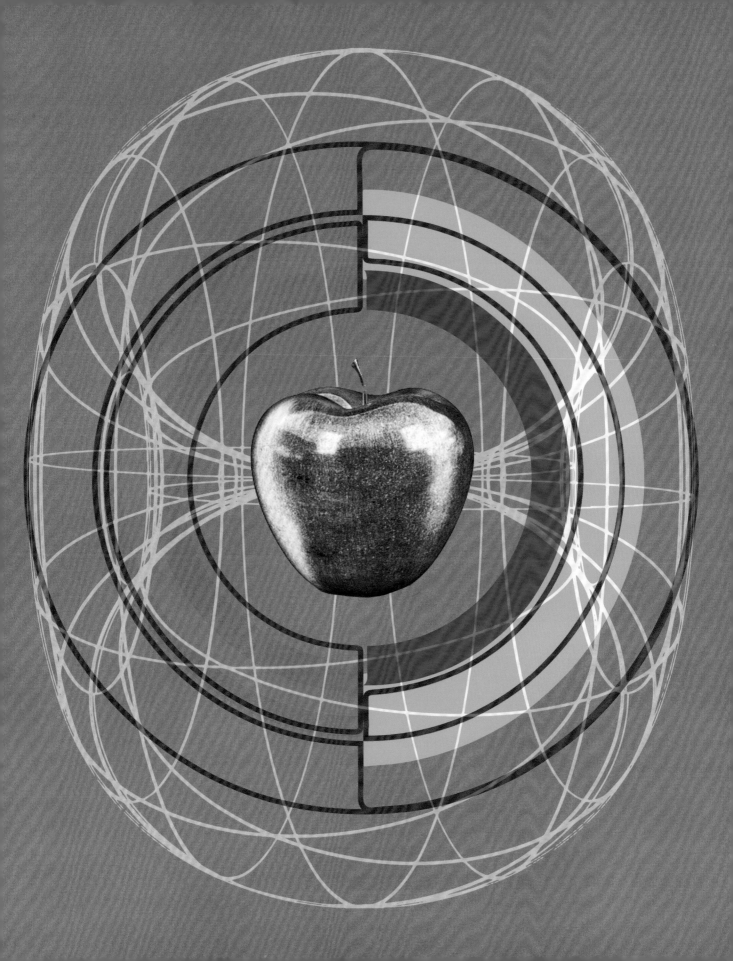

While the 1960s jazz-rock group Blood, Sweat & Tears made a convincing point when they suggested that what goes up must come down, they may have been more accurate by singing 'what possesses mass must distort space-time', but I think we can all understand why they opted for the former, somewhat catchier, lyric.

Apple-appraising academic Sir Isaac Newton was the first person to present a robust description of how gravity works through his laws of universal gravitation, presented in his 1687 work *Philosophiae Naturalis Principia Mathematica (Mathematical Principles of Natural Philosophy).*[1] It is thought to be one of the most important scientific books ever written, with Einstein describing it as 'perhaps the greatest intellectual stride that it has ever been granted to any man to make.'[2] While most of us, upon dropping a piece of fruit, ponder whether the fruit is still edible, when Newton observed a falling apple, he was struck by a different train of thought. Newton proposed that everything in the Universe that has mass exerts a gravitational 'pull' on every other object that has mass. His theory also proposed that the more 'massive' something is, the greater the gravitational force it exerts, but this force is dependent on another object's proximity to it.[3]

Newton's simple equations also explained why the Moon wouldn't crash into Earth like an apple falling to the ground. While the Moon is indeed 'falling' towards the Earth, it is moving in a curved trajectory and the Earth is curving away underneath at the same rate. This is the same for the International Space Station (ISS) and the astronauts inside. It is often described as zero-G in space, but the gravity experienced by astronauts on the ISS is actually 90 per cent of that on Earth. So why do astronauts float? It's because, like the Moon, the ISS is falling towards Earth, but it is also moving with such velocity that the Earth is moving away from it as it falls, so it misses, leaving the astronauts in a constant state of freefall. This allows the astronauts to enjoy the experience of weightlessness.[4]

Newton's laws of universal gravitation were unchallenged for hundreds of years until Albert Einstein offered an alternative view on gravity as part of his general theory of relativity, published in 1915.[5] Einstein proposed that gravity isn't a force where objects 'pull' on one another, but instead massive objects warp space-time and then other objects are influenced by this warping. In a rather brilliant YouTube video that has now achieved over 130 million views, a physics teacher called Dan Burns helped students visualise gravity with a large piece of Lycra and a series of different-sized marbles. The teacher placed a large marble at the centre of the stretched Lycra, causing a dip at the centre. When the teacher introduced another, smaller marble to the Lycra in a circular rolling fashion, it would circle the centre, following the central dip caused by the larger marble.[6] This is a simplified, but wonderfully impactful way to help understand the distortion of space-time.

In addition to the distortion of space-time, Einstein's general theory of relativity has predicted numerous phenomena, such as black holes (matter becoming so dense and distorting space-time so much that not even light 'escapes' it), gravitational waves (the ripples in the very fabric of space-time as a result of collisions between bodies of enormous mass, such as two black holes) and gravitational lensing (light following the curvature of space-time around massive objects, which Newton's explanation of gravity as a force between objects with mass wouldn't explain, considering light particles, known as photons, have no mass).[7]

Perhaps one of the most fascinating elements of the distortion of space-time is gravitational time dilation and its impact on the experience of time by the observer. The closer an observer is to an object of huge mass, the slower the experience of time. Even GPS satellites orbiting Earth are calibrated to account for the fact that time is running more slowly on Earth than it is in orbit, albeit by a very, very small amount. Simply put, the greater the mass, the greater the experience of time dilation to the nearby observer.

Being a sucker for operatic science-fiction cinema, I'm a huge fan of Christopher Nolan's *Interstellar* – a film whose scientific consultant was eminent astrophysicist Kip Thorne – which contains a scene that creatively portrays time dilation. This scene witnesses the film's lead character trapped on a planet situated so close to a black hole that during the three hours spent there, twenty-three years have passed on Earth. A fact devastatingly illustrated when the protagonist returns to his spacecraft to watch two decades' worth of video messages from his daughter. While this is clearly an extreme example that has admittedly been creatively massaged for our cinematic enjoyment, the complicated science of time dilation is communicated simply and to powerful effect. While a great deal of artistic licence is necessary in cinema, Christopher Nolan and Kip Thorne went to great lengths to ensure that where possible, *Interstellar* was grounded in science. Thorne's book *The Science of Interstellar*[8] showed how many of the ideas in the film were born from scientific concepts. In an interview with *Scientific American*'s Lee Billings, Thorne said, 'Real science can give rise to wonderful ideas for a film that can in most cases be better than what was created from whole cloth out of the brain of a screenwriter ... this is the first film where that

rule set closely corresponds to the known laws of nature, and some truly wonderful things came out of this. Real science can be an absolutely fabulous foundation for great filmmaking.'[9]

But even with the advancements in understanding made by intellectual titans such as Matthew McConaughey, Newton and Einstein, our comprehension of gravity is still incomplete. While these theories have enabled us to understand the movements of planets, solar systems and galaxies, we are still unable to reconcile the behaviour of the very big with that of the very small. Things that exist at a subatomic level simply do not comply with our current understanding of physics and require a different set of rules. This is why scientists are searching for a grand unified theory or theory of everything that reconciles it all into one coherent model of classical and quantum physics.[10]

Sir Isaac Newton said, 'If I have seen further, it is by standing on the shoulders of giants,'[11] and with astrophysics there is certainly an abundance of intellectual giants who have furthered our journey towards understanding gravity. But this journey is clearly far from over. In addition to the search for a unifying theory that reconciles general relativity with quantum mechanics (see 'Q is for Quantum Physics'), the lack of knowledge surrounding the accelerating expansion of the Universe may also mean that we have flaws in our understanding of gravity on a cosmological scale. Einstein said, 'As our circle of knowledge expands, so does the circumference of darkness surrounding it,'[12] suggesting that as we solve one puzzle, more will reveal themselves.

Q

Where do theoretical physicists go to find love?

A

Singularities night.

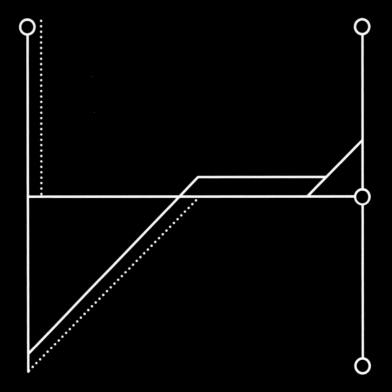

IS FOR

Hubble

Around 550 km above Earth, the world's first space telescope orbits our planet at over 27,000 km an hour, peering deep into the cosmos. An omniscient eye staring into the dark. This school-bus-sized space observatory has offered unparalleled views of distant galaxies that have truly transformed our understanding of the Universe.

The Hubble telescope was launched in 1990, named after Edwin Hubble, one of the people credited with discovering that the Universe is expanding.[1] In 1969, the National Academy of Sciences published *Scientific Uses of the Large Telescope*, which argued the important benefits of a non-terrestrial telescope in furthering our understanding of the cosmos.[2] Earth-based telescopes suffer from a number of limitations due to the Earth's atmosphere, which distorts and absorbs the light coming from stars. Any telescope that could escape these limitations would offer profound advancements in our ability to study the Universe.

The pioneering space observatory simply wouldn't have happened if it weren't for the tenacity and hard work of one extraordinary woman known as the 'Mother of Hubble'. It was astronomer and NASA's first chief of astronomy, Nancy Grace Roman, who convinced Congress that the launch of a powerful space telescope was of vital importance and that for the price of a cinema ticket, each American could be given years of scientific discoveries.[3]

Through international co-operation from NASA and the European Space Agency to reduce the overall costs, the proposal to build the telescope was approved in 1977, but numerous delays meant the launch didn't take place for another thirteen years, by which point the expectations surrounding which images Hubble might capture were quite literally astronomical.

While Hubble is undoubtedly one of science's most incredible achievements, there was a period after its 1990 launch when the telescope was nearly an unmitigated disaster. After billions of dollars in investment and over a decade in planning, when Hubble reached orbit and began capturing images of the Universe, the first to return to Earth were blurry. 'Blurry pictures' and 'multibillion-dollar lens' are two phrases that do not go very well together. Understandably, this was at the very best disappointing, and the worst, utterly devastating. It was discovered that Hubble had a flaw in its primary mirror that was affecting the clarity of its images – essentially, the mirror was the wrong shape, rendering Hubble unable to focus correctly.[4] While the misalignment was only about 1 mm, this was a huge problem for Hubble's efficacy. Fortunately, Hubble was the first telescope designed to be serviced in space by astronauts, so the fault had the potential to be rectified. Optical engineers and scientists from both the US and Europe set about designing a solution to get Hubble back on track.

When Hubble was due its first servicing mission in 1993, astronauts added a corrective system that essentially acted as a pair of glasses,[5] precisely counteracting the misshapen primary mirror. Fortunately, social media wasn't a thing in the 1990s, otherwise Specsavers' social media team would have constructed a twelve-month content plan around this one mistake. The servicing mission was a resounding success, and the new, sharpened images were nothing short of spectacular, offering a wonderful glimpse of what Hubble was going to achieve in its lifetime.

According to NASA:

The orbiting telescope has taken over a million observations and provided data that astronomers have used to write more than 14,000 peer-reviewed scientific publications on a broad range of topics, from planet formation to gigantic black holes. These papers have been referenced in other publications over 600,000 times, and this total increases, on average, by more than 150 per day. Every current astronomy textbook includes contributions from the observatory. Today's college undergraduates have not known a time in their lives when astronomers were not actively making discoveries with Hubble data. [6]

Hubble's contract was extended for another five years in 2016, which saw it overlap with the launch of another marvel of engineering, NASA's James Webb Space Telescope (JWST), which has now become the primary observatory and will continue Hubble's legacy in space. The JWST's improved technology (such as longer wavelength coverage and advanced sensitivity) allows it to look much more closely at the beginning of time and to search for the unobserved formation of the first galaxies.[7] The JWST's ability to observe the light omitted from these early stages of the Universe means it behaves as a time machine, searching into the history of the cosmos. NASA administrator Bill Nelson said, 'The James Webb Space Telescope represents the ambition that NASA and our partners maintain to propel us forward into the future. The promise of Webb is not what we know we will discover; it's what we don't yet understand or can't yet fathom about our universe. I can't wait to see what it uncovers!'[8] At the time of completing this book, breathtaking images were starting to arrive from the JWST, adorning news outlets all over the world and hopefully inspiring new minds to contemplate the Universe.

Hubble has aided numerous groundbreaking developments in our understanding of the mechanics of the Universe. Just some of these extraordinary contributions include the census that led to the theory that all galaxies have a supermassive black hole at the centre; the observations of supernovae that suggest the Universe's expansion is accelerating; and the measuring of distant galaxies and the oldest star clusters to pinpoint the age of the Universe.[9]

Since Hubble began its mission over thirty years ago, there have been countless images that astound, not just from a scientific perspective but because of these images' sheer, unadulterated beauty.[10] From revealing the machinations of exploding supernovae, the formation of stars, gravitational lensing around galaxy clusters and looking billions of years into the past, these astounding images have continued to capture the public's imagination, inspiring new generations of astronomers and furthering the accessibility and communication of science to a broad audience. I often take solace that even behind the dull yellow canopy of central London's night sky, there is an infinity of explosions, whirlpools and cascading spirals of colour in every direction, and we are lucky enough to be able to witness just some of this beauty through the ingenuity and dedication of those who made and continue to make space observation possible.

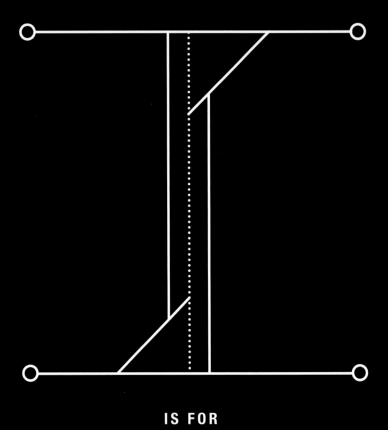

IS FOR

International
Space Station

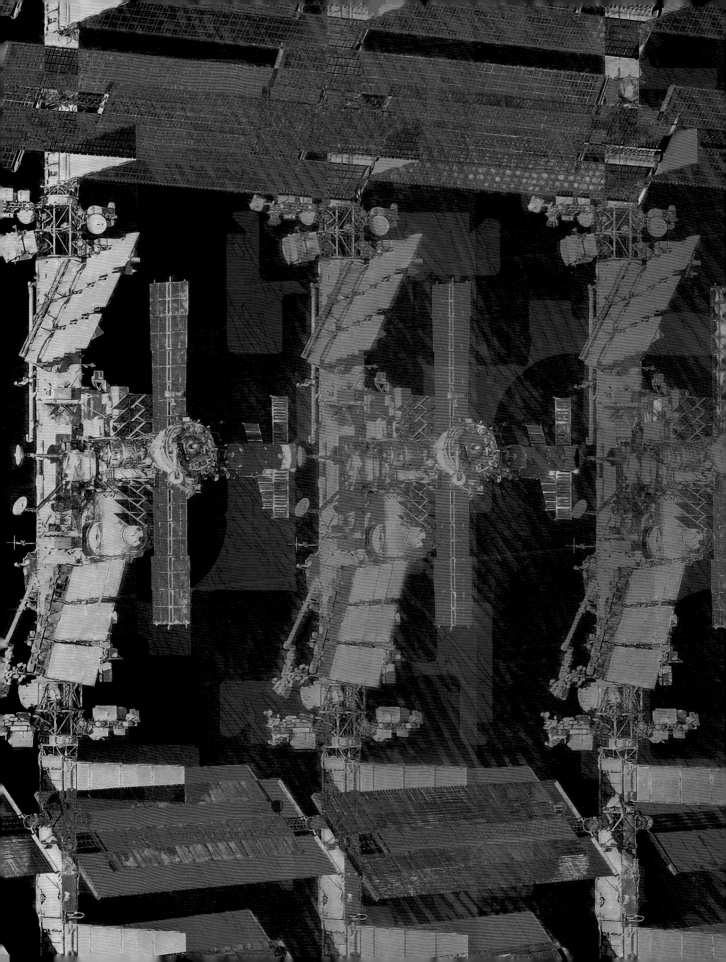

Despite his famous song being released the same year as the Moon landing, it's unlikely that David Bowie would have envisaged an astronaut by the name of Colonel Chris Hadfield performing and recording a music video for 'Space Oddity' on the International Space Station (ISS) over forty years later. Just a few years before Bowie's death, his official Facebook page described the recording as 'possibly the most poignant version of the song ever created'.[1] Seeing Chris Hadfield deliver Bowie's lyrics while gazing upon our planet as it spins 400 km beneath, it's easy to see why.

Cruising at a speed of over 27,000 kph, the ISS orbits the Earth every ninety minutes, allowing it to witness sixteen sunsets every single day. Being around the same length as a football pitch helps the ISS to be one of the most visible objects in the night sky and it can be easily seen on a clear evening, shooting across the sky like a comet.

To achieve something so astonishingly ambitious as a permanently staffed space laboratory was a truly global effort symbolised by the ISS Intergovernmental Agreement (IGA). This treaty was signed by the United States, Canada, Japan and the ten founding European countries of the European Space Agency, setting the legal framework surrounding the ISS.[2]

The first segment of the ISS was launched on 20 November 1998, and within just two years, two Russian cosmonauts and one American astronaut became the first crew to live onboard. Since this first crew arrived, the ISS has been staffed permanently, with over 240 individuals having visited from nineteen different countries.[3]

The primary function of the ISS is to serve as a space laboratory that furthers our understanding of what it takes to live outside the safety of our own planet. One of the main benefits of the ISS is the appearance of weightlessness. As we've touched on in 'G is for

Gravity', contrary to what many think, astronauts inside the ISS are not actually operating in zero gravity – gravity on the ISS is actually around 90 per cent of that experienced on Earth.[4] In reality, the ISS (and the astronauts inside it) are in perpetual freefall around Earth – the speed at which the ISS is 'falling' is matched by the curvature of Earth's surface, leading to the appearance of weightlessness on board. This allows the astronauts to conduct their experiments within a microgravity environment and begin to understand how humans might live for prolonged periods in space.

While on board the ISS, astronauts undertake numerous duties, including implementing research into technologies that will be required for future space exploration. Astronauts also take part in medical experiments to observe the effects of microgravity on the human body, including the loss of bone density, decrease in vision and reduced pressure on the cardiovascular system. It's vital for astronauts to exercise daily in space to counter these effects and avoid long-term damage.[5] And the research into biology on board the ISS goes even deeper. According to NASA, 'controls on the directionality and geometry of cell and tissue growth can be dramatically different to those on Earth', so astronauts also conduct experiments using the culture of cells, tissues and small organisms[6] to further our understanding of the biological processes in microgravity.

Not only is the data gathered on the ISS of benefit to the future exploration of the Universe, it also provides a wealth of information and innovation that impacts life on Earth. Such benefits include the monitoring of natural disasters, the gathering of data on coral reefs, agriculture and glaciers, and the development of water-filtration technology that has impacted water purification worldwide,[7] to name just a few.

Due to the astronauts' workplace and living space being one and the same for months at a time, the ISS has to enable residents to attend to their daily needs while minimising the risk of various errant fluids. However, the manner in which they have to attend to these basic needs must take into account that they are travelling at 27,000 kph while experiencing microgravity. Let's just say there's very little debate on the ISS as to who left the toilet seat up.

One of the main problems with the experience of microgravity inside the ISS is that everything floats, not just the astronauts, so this needs to be borne in mind with personal hygiene. Despite the ISS itself being one of the most elaborate machines ever created, the system of sanitary ergonomics is surprisingly simple. Zero-gravity toilets use the art of suction to unceremoniously divorce an astronaut from their waste. United States astronaut Mike Massimino revealed that prior to visiting the ISS, astronauts hone rather sophisticated target practice using a dummy ISS toilet bowl. If this doesn't sound awkward enough already, there's the addition of cameras to ensure that their aim is accurate.[8] Let's hope all video evidence of such target practice has been destroyed.

Taking a shower in the traditional sense is also something that is simply not possible on the ISS. With no bathtub to rub-a-dub in or shower to sing under, astronauts simply apply water directly to their skin from a water pouch and, due to surface tension, it remains on the skin, where they can also apply soap. They then wipe away with a flannel.[9]

Getting anything off Earth and into space is extremely difficult, so the cost of transporting something to the ISS is quite literally astronomical. The astronauts' food has to tread a fine balance between being tasty, of high nutritional value, able to last for a long time without going off, and also weigh as little as possible to keep costs down, as it's around $10,000 per 450 g serving.[10] It's these huge costs that also necessitate the ISS's water-recovery system, which turns urine and wastewater into clean water. As succinctly expressed by Koichi Wakata of the Japan Aerospace Exploration Agency, 'On board the ISS, we turn yesterday's coffee into tomorrow's coffee.'[11] Not a slogan you'll find on a ground-coffee pack in the supermarket any time soon.

If the thought of a $1,000 Mars bar or recycled urine doesn't make you break out into a cold sweat, then perhaps you might be interested in being one of the first private astronauts to pay a visit to the ISS? It was recently announced that the first commercial crew would be flying to the International Space Station in the coming years. The private company Axiom Space will send its crew aboard a SpaceX craft and will reportedly be paying $55 million for each crew member.[12] Axiom is also building the world's first private space station in Earth's orbit, described on their website as 'a commercial laboratory and residential infrastructure in space that will serve as a home to microgravity experiments'.[13] With unobstructed 24/7 views of Planet Earth from low orbit while experiencing microgravity, it gives a whole new meaning to Airbnb.

But what of the future of the ISS? Well, after circling the planet since 1998, its destiny lies at the bottom of the Pacific Ocean. NASA announced in 2022 that the ISS would be taken out of orbit in 2031 by crashing into a place known as the 'spacecraft cemetery'.[14] The controlled descent will involve the ISS gradually lowering its orbit into an increasingly dense atmosphere, slowing the craft down, with it eventually falling into an uninhabited area in the middle of the Pacific Ocean. With the ISS taking residence deep in the ocean, future astronauts will be reliant on commercial space stations to continue their work in the skies.[15]

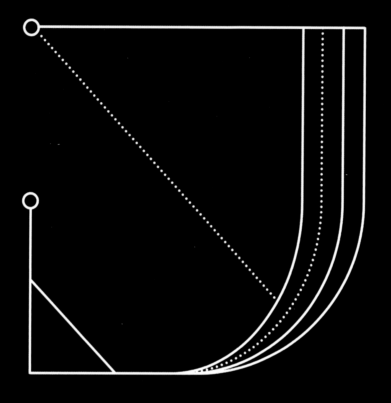

IS FOR

Jupiter

The use of the word 'big' in common vernacular such as 'big deal', 'big shot' and 'Big Mac' has somewhat left us without appropriate words to describe just how big something like Jupiter really is. Jupiter is so big you could fit Earth inside it over 1,300 times.[1] Jupiter is so big that if it were only eighty times more massive, it may have kick-started thermonuclear reactions and become a star.[2] It's so big that it's visible to the naked eye and is the third brightest object in the night sky despite being almost 590 million km from Earth.[3] If you were still in any doubt as to how big Jupiter is, you could happily fit all the planets of the solar system comfortably inside it.[4]

So, we've established that the word *big* doesn't do Jupiter justice, but this is perhaps no better observed than in its weighty influence upon our very own Sun. Jupiter, being twice as massive as all the planets combined, exerts such a significant gravitational influence that it even makes the Sun wobble.[5] This is due to an area called a barycenter – the centre of mass for the solar system. Common knowledge would assume that this would be dead centre within the Sun, but the Sun's aforementioned wobble is due to the gravitational effects of all the planets in the solar system, so the actual barycenter shifts depending on the position of the planets in orbit. Sometimes it's close to the Sun's centre, sometimes it's just outside the surface.

Being a very large planet composed mostly of hydrogen and helium, Jupiter is known as a gas giant. Like the other gas giants in our solar system, Neptune, Saturn and Uranus, Jupiter possesses a largely gas exterior with what is thought to be a small rocky core.[6] Being formed from much of the leftovers from the Sun, Jupiter has similar ingredients to a star, but didn't have the requisite mass to become a star in its own right. Must be tough being a failed star and constantly gassy, but despite its size, it rotates much more quickly than Earth, with one of its days lasting just under ten hours. However, its journey around the Sun is much more relaxed than ours, taking around twelve years for a full solar lap.

Ancient tablets reveal that the path of Jupiter was recorded first by the Babylonians 2,400 years ago using a form of calculus that was previously thought to have been developed in the fourteenth century.[7] Jupiter was given its name by the Romans, who named it after their god of sky and thunder. Thunder, in fact, is a rather appropriate reference, due to Jupiter's famous Great Red Spot – a giant storm which may have raged for as long as 350 years, perhaps even longer. This spot, perhaps Jupiter's most prominent feature, is so large it could fit three Earths inside it. A lesser-known feature of Jupiter is its Saturn-like rings. However, where Saturn's rings are clearly visible, due to their high quantity of ice, making them reflective, Jupiter's are mainly composed of dust, rendering them largely invisible.[8]

It's been theorised by astronomers that we should be thankful for sharing a solar system with Jupiter, its enormous gravitational strength attracting comets and space debris that may otherwise have collided with Earth. But more recently it's been suggested that Jupiter hasn't been quite the cosmic protector we've previously thought. According to planetary scientist Kevin Grazier, Jupiter may be a 'sniper rather a shield', flinging unwanted debris in our direction.[9] So perhaps we shouldn't offer our endless fealty to the giant just yet.

Unsurprisingly for its size and gravitational heft, Jupiter has a large number of moons circling it. So many moons, in fact, that we've only found time to name fifty-three of them, with a further twenty-six still begging to be baptised.[10] Despite being in the shadow of the solar system's biggest planet, some of these less conspicuous moons are just as fascinating as the behemoth itself. The largest of them – Io, Europa, Callisto and Ganymede – are known as the Galilean moons because they were discovered by – guess who? – Galileo in 1610. Each of these hold distinctive and captivating characteristics, but two in particular caught my attention – one eternally burning like Hades manifest and another that potentially harbours an incredible secret.

Io is covered in over 400 volcanoes, making it the most volcanically active body in the solar system. Sounding like a doomsday prophecy, its surface is covered in sulphur and lava lakes and its impressive volcanic plumes explode high above the surface.[11] It was one of these ejections, spotted in an image by the *Voyager* spacecraft, that revealed it wasn't in fact a dead moon, but the hellish, tempestuous object we know today. The reason for Io's infernal personality is the strong gravitational pull from Jupiter on one side and the pull of the other Galilean moons on the other. This cosmic tug of war causes tidal bulges on the moon's surface, stretching and heating Io in the process. Considering this is solid ground, it's amazing to envisage these enormous bulges reaching as high as 100 m above the surface.[12]

In contrast to Io's blazing surface, Europa is covered with an ice crust – this, intriguingly, has led scientists to believe there may be an ocean of water underneath.[13] Much like Io, Europa experiences tidal flexing from Jupiter's gravitational force, which is believed to be causing the heating of Europa's interior and therefore the possibility of maintaining an ocean of liquid. The cracks on the icy surface also imply tidal movement underneath the ice, causing the surface to rise and fall. Images from the Hubble

Space Telescope revealed that Europa might be ejecting plumes of water above the surface which would, in turn, imply geological activity. This could position Europa as the most promising place for life in our solar system after Earth. Research has shown that volcanic activity may have taken place on Europa's ocean floor, which could offer even further indication that life is possible under the icy crust. Marie Běhounková of Charles University, who led the research, said to NASA, 'Our findings provide additional evidence that Europa's subsurface ocean may be an environment suitable for the emergence of life. Europa is one of the rare planetary bodies that might have maintained volcanic activity over billions of years, and possibly the only one beyond Earth that has large water reservoirs and a long-lived source of energy.'[14]

NASA is due to launch the spacecraft *Clipper* in 2024, which will see the craft undertake frequent fly-bys of Europa to assess its suitability for life.[15] Robert Pappalardo, Europa Clipper project scientist, said, 'The prospect for a hot, rocky interior, and volcanoes on Europa's seafloor, increases the chance that Europa's ocean could be a habitable environment. We may be able to test this with Europa *Clipper*'s planned gravity and compositional measurements, which is an exciting prospect.'[16] While the Clipper itself may not fully reveal exactly what lies beneath the ice, it seems to be just a matter of time until a spacecraft is sent which can uncover the mystery.

'Going to the Kuiper Belt is like an archaeological dig into the history of the solar system.'

ALAN STERN

PRINCIPAL INVESTIGATOR, NEW HORIZONS

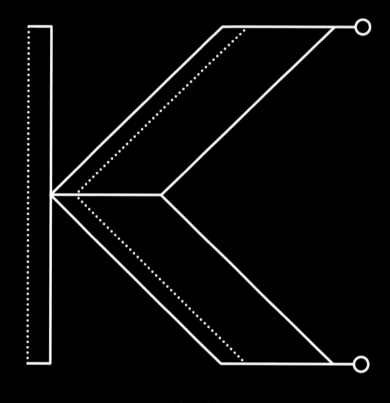

IS FOR

Kuiper Belt

If you left Earth and travelled around 4.5 billion km away from the Sun, you would reach the inner ring of the Kuiper Belt.[1] To put this in context, if you took the average family car and drove as fast as you could in the Kuiper Belt's general direction, you'd reach its inner ring somewhere around the year 4500, give or take the occasional toilet break. Perhaps Elon Musk had the same idea when he sent his Tesla towards Mars. Despite these gargantuan numbers, on a cosmic scale, the Kuiper Belt is very much on our doorstep – but it's a region of our solar system that we've only recently been able to get closer to.

Despite being referred to as a *belt*, this structure within the solar system is in fact more akin to a doughnut in dimensions. It begins at Neptune's orbit, which is around thirty astronomical units from the Sun, and ends at a region around fifty astronomical units from the Sun. To help fathom the large distances involved, one astronomical unit is around 150 million km – the distance between the Sun and Planet Earth. The Kuiper Belt is made up largely of icy objects, comets and even dwarf planets that astronomers think are the leftovers from the formation of the solar system. These may well have formed to create their own planet if it weren't for the gravitational effects of Neptune preventing them from joining in cosmic matrimony.[2]

As with many scientific progressions, the realisation that the Kuiper Belt existed wasn't the product of a single discovery but a succession of theories and observations from multiple people. Shortly after the discovery of Pluto in 1930 by Clyde Tombaugh, an American astronomer at the University of California called Frederick C. Leonard believed that Pluto might be the first in a number of objects past Neptune that were waiting to be discovered. However, the Kuiper Belt is named after the Dutch American astronomer

Gerard Kuiper who, in 1951, published a paper that speculated about objects beyond Neptune and a disc that may have formed in the early evolution of the solar system.[3] Leonard wasn't the first person to put forward similar theories prior to Kuiper, which has led to some contention around its name. Prior to Kuiper's theories, Irish astronomer Kenneth Edgeworth had contemplated that Pluto was in fact too small to be a planet and might well be part of a larger zone encompassing smaller bodies.[4] With a number of different people hypothesising about this part of the solar system, it is understandable that some may object to Kuiper getting so much of the glory. This has led to some astronomers referring to this zone as the Edgeworth-Kuiper Belt to acknowledge some of the different contributions. So, the next time you find yourself propping up the bar deep in conversation with a number of astronomers and observational cosmologists, and you're feeling diplomatic, perhaps you might consider using the term 'Edgeworth-Kuiper-Leonard-Doughnut Belt' to hedge your bets.[5]

The Kuiper Belt is also thought to be the home of another cosmic phenomenon that must have caused puzzlement and inspired many a wish through the millennia. Comets are typically composed of ice and dust, are irregularly shaped, and measure anything from hundreds to thousands of metres across. Those originating in the Kuiper Belt are known as 'short period' comets, which orbit the Sun in under 200 years, performing a slingshot action around it before travelling out past Neptune. These differ from 'long period' comets, which are thought to originate outside the Kuiper Belt in the Oort Cloud and can take thousands of years to complete their elongated, lonely journeys to the far reaches of the solar system and back.[6]

Despite the multitude of hypotheses surrounding this part of our cosmological neighbourhood, the Kuiper Belt wasn't discovered until over forty years after Kuiper's paper. In the summer of 1992, English astronomer David Jewitt and Vietnamese American astronomer Jane Luu discovered the very first Kuiper Belt object (other than Pluto and its moon, Charon), catchily named '(15760) 1992 QB1'.[7] Due to the significance of the discovery, it comes as quite a surprise to me that the name (15760) 1992 QB1 didn't make it onto the most popular baby names list of 1992, alongside Harry and Sophia.

Since the discovery of (15760) 1992 QB1 (now known as the marginally more memorable 15760 Albion), over 2,000 other Kuiper Belt objects have been documented by observers,[8] and it's even been estimated by astronomers that there may be hundreds of thousands more objects of at least 100 km across.[9] In addition to the five confirmed dwarf planets – Ceres, Eris, Makemake, Haumea and the most famous, Pluto – there are other objects that have been observed that could also qualify as dwarf planets, but due to their size and distance from Earth have yet to be confirmed.

When observing something as far away as the Kuiper Belt, there are obvious limitations when using equipment on Earth, which is why *New Horizons*, the first spacecraft to visit this region of the solar system, was so exciting. At the time of launch, *New Horizons* project manager Glen Fountain said, 'This is the gateway to a long, exciting journey ... The team has worked hard for the past four years to get the spacecraft ready for the voyage to Pluto and beyond, to places we've never seen up close. This is a once-in-a-lifetime opportunity, in the tradition of the *Mariner, Pioneer* and *Voyager* missions, to set out for first looks in our solar system.'[10]

New Horizons left Earth in 2006, passed Jupiter a year later and reached Pluto in 2015. The spacecraft captured spectacular images of Pluto's purple-and-white icy surface, dramatic mountain ranges and deep craters before continuing deeper into the Kuiper Belt – further than any human-built object has ever travelled before. A journey to the darkest, coldest reaches of the solar system. *New Horizons* is currently almost 8 billion km from Earth and will continue its journey through the Kuiper Belt, eventually leaving the solar system entirely.[11]

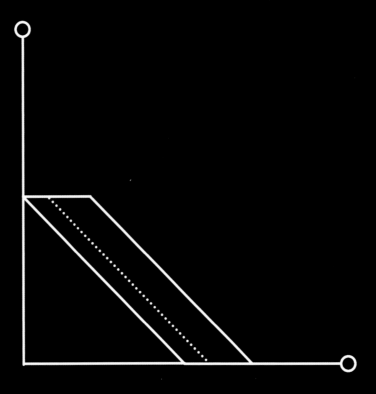

IS FOR

Large Hadron Collider

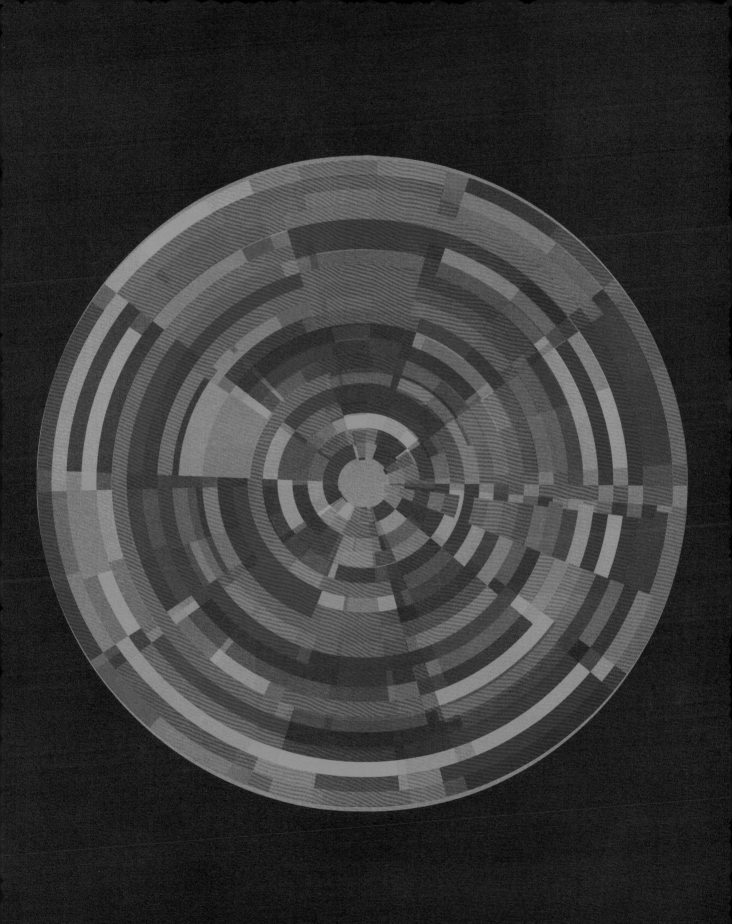

Picture this: one of humanity's most incredible feats of science is about to be switched on. Thousands of the planet's most extraordinary brains have worked for years to develop a machine that may answer some of science's most difficult questions, potentially shining a light on the very fabric of existence. The world waits with bated breath, poised and ready to see what wonders will reveal themselves... Scientists the world over hold a collective breath of anticipation, and one newspaper asks, 'Are we all going to die next Wednesday?'[1]

The idea of a subterranean tunnel designed to purposefully crash things into each other at high speeds was clearly enough to get people twitchy. Especially given that the Large Hadron Collider (LHC) was rumoured to be producing mini black holes and recreating the conditions of the big bang. The internet was flooded with existential questions, not just surrounding the LHC's launch but also in the years that followed. In 2019 an astronomer was quoted as saying, 'The Large Hadron Collier [sic] could create BLACK HOLE and DESTROY EARTH'[2] and another as saying, 'Earth could be reduced to 330 ft sphere if Large Hadron Collider experiments go wrong.'[3] Even CERN (the European Organization for Nuclear Research and home of the Large Hadron Collider) posted an article on their official website hoping to allay any concerns around the potential for rogue black holes.[4] Fortunately for us, when the Large Hadron Collider was formally tested in 2008, Earth wasn't reduced to the size of a football pitch and existence as we know it didn't come to an end.

In a 2009 TED Talk, Professor Brian Cox, physicist and researcher at CERN, said that the machine aims to 'recreate the conditions that were present less than a billionth of a second after the Universe began, up to 600 times a second'.[5] The LHC is the world's largest particle accelerator, and recreating these moments directly after the big bang will further our understanding of the fabric of the Universe and hopefully answer many questions that are not answered by the standard model of physics.

But large ambitions invariably require a rather large machine. Just a short distance west of Lake Geneva, and over 100 m underground, lies a 27-km-long circular tunnel that houses a ring of superconducting magnets that accelerate particles close to the speed of light. Two particle beams are hurtled in opposite directions at 99.9999 times the speed of light to create collisions between particles. Scientists at CERN then observe the byproducts of these impacts to gain a deeper understanding of the behaviour of the subatomic Universe – the very building blocks of our existence.

The LHC is just one part of CERN's multifaceted research. Founded in 1954, there is no better testament to the unifying abilities of the quest for discovery than this organisation – a place where scientists from all over the world unite for the furtherment of science. Some of the discoveries and accolades associated with CERN prior to the launch of the LHC include the discovery of elementary 'weak force' W and Z particles in 1983, leading to a Nobel Prize in 1984,[6] and the birth of the World Wide Web in 1989. Tim Berners-Lee was a physicist at CERN at the time, and the Web was designed as an information-sharing platform for scientists around the world.[7]

While the LHC was first tested in 2008, according to physics professor Chris Llewellyn Smith in his article 'Genesis of the Large Hadron Collider', the planning started in 1976.[8] The creation of such a monumental machine took the work of 10,000 scientists, $4.75 billion and fourteen years to assemble.[9] It's not hyperbole to say that it's one of the most extraordinary feats of science and engineering ever known to humanity, especially as one of its first high-profile and pioneering discoveries came within just four years of its activation.

Forty-eight years prior to its official discovery at the Large Hadron Collider, Peter Higgs had theorised about a particle that helped give other particles mass – a particle associated with the Higgs field that flowed through all of space-time. Until experiments began at the LHC, the existence of such a particle was just part of the theory of the standard model of physics but had not been observed. On 4 July 2012, scientists at CERN announced that they had discovered a particle consistent with the long-sought Higgs boson. The following year, Peter Higgs and François Englert were awarded the Nobel Prize in Physics for 'the theoretical discovery of a mechanism that contributes to our understanding of the origin of mass of subatomic particles, and which recently was confirmed through the discovery of the predicted fundamental particle, by the ATLAS and CMS experiments at CERN's Large Hadron Collider'.[10]

But what does the discovery of the Higgs boson, or to be more precise, the discovery of a 'Higgs-boson-like' particle mean for our future understanding of the Universe? According to CERN's official website, 'the Higgs boson holds the key to our understanding of nature beyond what is shown by the Standard Model'.[11] By observing these particles, scientists at CERN theorise that the Higgs boson may transform into undetectable particles which could comprise the ubiquitous dark matter – the currently unknown particles that are thought to account for 85 per cent of the Universe's matter but are, as yet, unobservable and undetectable.

But even $4.75-billion pioneering scientific-marvel subterranean super colliders need to hang up their weary superconducting magnets at some point. When the Large Hadron Collider reaches the end of its lifespan in the coming years, there are plans to replace it with an even larger machine, its details laid out in the Future Circular Collider (FCC) study at CERN. The Future Circular Collider would involve a whopping 100-km tunnel that will enable scientists to observe collision with far more accuracy than currently possible with the LHC.[12]

The £20 billion price tag for such a massive collider has understandably raised some concerns as to the cost benefit of such experiments. Sabine Hossenfelder, a theoretical physicist, commented, 'Another reason I am not excited about the current plans for a larger collider is that we might get more bang for the buck if we waited for better technologies ... we do not currently have a reason to think it would discover anything new, i.e. large cost, little benefit.'[13] Defenders of future colliders would argue that it is impossible to judge the potential rewards in the short term, as such innovative endeavours invariably will only reveal their true benefits – in our understanding of the Universe and its practical implications – years, decades, perhaps centuries down the line.

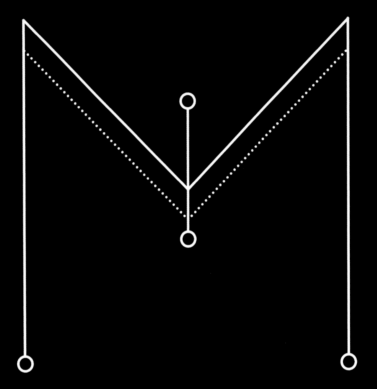

IS FOR

Milky Way

It is estimated that there could be around 2 trillion galaxies in the observable Universe sprawling across the night sky,[1] of many different shapes and sizes. Our own planet currently sits in between the two primary arms of a spiral, disc-shaped galaxy called the Milky Way. To get into spiral-galaxy semantics (if you'll indulge me), our galaxy is technically a 'barred' spiral galaxy, meaning the arms extend from the ends of a bar shape at the core, as opposed to a standard spiral galaxy, where the arms curve outwards from a circular core.[2]

Our galaxy is called the Milky Way due to its milky appearance, although from our perspective on one of the galaxy's arms, we're only seeing small portions of it stretching across the sky at any one time. As with most scientific achievements, our understanding of the Milky Way, and galaxies in general, has broadened through incremental discoveries over the years, so here's a whistle-stop highlights reel of some of the key moments:

The name Milky Way comes from the ancient Greek *galaktikos kyklos*, meaning 'milky circle', and it was this milky circle that inspired ancient Greek philosopher Democritus to ponder whether it could be the light from distant stars.[3] This was a pretty radical opinion at the time, as the Milky Way was more associated with being a manifestation of the gods in the sky. But it was Galileo Galilei who became the first to see the individual stars of the Milky Way in 1610, describing 'congeries of innumerable stars grouped together in clusters too small and distant to be resolved into individual stars by the naked eye'.[4] In 1918, American astronomer Harlow Shapley established that the centre of our galaxy was in the constellation of Sagittarius.[5] It was Edwin Hubble in the 1920s who first provided conclusive proof that

the Andromeda Nebula was a separate galaxy and went on to show that our own galaxy is just one of many others, all moving away from one another as a result of the ongoing expansion of the Universe.[6] Understanding the nature of the Milky Way has definitely been a multigenerational effort.

Planet Earth is located around 27,000 light years from the Milky Way's centre, on the inner curve of the Orion Arm,[7] and while we travel around the Sun, the Sun is also making its way around the Milky Way. However, the galaxy is so large that the last time our planet saw the other side of the Milky Way, dinosaurs were comfortably ruling the planet and were midway through the Cretaceous period.[8] It seems a great deal can happen in the 220-million-year journey around the centre of the galaxy, including the demise of the dinosaurs and emergence of the human race.

By looking at the behaviour of stars and matter near the centre of galaxies, astronomers have theorised that most possess a supermassive black hole at the nucleus, and the Milky Way is no different. Going by the name Sagittarius A*, the supermassive black hole thought to be at the Milky Way's centre has a mass 4 million times that of our Sun[9] and its influence may have been vital in the formation of the whole galaxy.[10]

By all accounts, other than being home to our planet, the Milky Way is a fairly average galaxy at around 100,000 light years across. Due to the density of matter in its centre, it's impossible for us to know exactly how many stars our galaxy contains. However, the Gaia space observatory, which launched in 2013, is helping to create a three-dimensional map of our galaxy, and its hugely ambitious star-scanning mission has already mapped around 1.7

billion stars.[11] As large as this number seems, it is only a small fraction of the total number of stars in our galaxy, with estimates ranging from 100 to 400 billion.[12]

Data from the Gaia satellite has shown that while disc-shaped, the Milky Way is not completely flat but warped, with a slight upwards curve on one side and a downwards curve on the other, moving similarly to a spinning top as it turns. This has led to astronomers believing that this warping could be the result of an ongoing collision with another galaxy. This collision, which could be with the nearby dwarf galaxy Sagittarius, sends waves through said galaxy, causing the observed warping.[13]

With the existence of potentially hundreds of billions of stars in the Milky Way, it begs the question: are there other habitable planets much like our own? Astronomers from the University of British Columbia say that not only are there many other Earth-like planets, there could be as many as 6 billion across our galaxy. To be considered Earth-like, planets must meet certain criteria such as being a similar size to Earth, rocky, orbiting a type of star similar to our own and orbiting at a similar distance within a habitable zone where the possibility for the development of life is optimal.[14] Six billion Earth-like planets in the Milky Way means 6 billion opportunities for life to be flourishing right now. How do these numbers translate to the probability of intelligent civilisations within the Milky Way? A paper published in *The Astronomical Journal*[15] described the possibilities for intelligent life within the Milky Way based specifically on conditions necessary for life to develop here on Earth — the one place we know for a fact that life has thrived. The paper asserts that, based on civilisation taking

somewhere between 4.5 and 5.5 billion years to evolve here on Earth and the number of estimated habitable Earth-like planets in the galaxy, there should be at least thirty-six communicating extraterrestrial intelligent civilisations. However, even if they were capable of communication, if these civilisations were spread evenly through the galaxy, the nearest one to us would be 17,000 light years away, rendering impossible any practical interplanetary communication as we might recognise it.

Perhaps for some, the knowledge that there may be intelligent life out there makes the night sky a little less empty. For now, even if we aren't alone, it certainly feels that way.

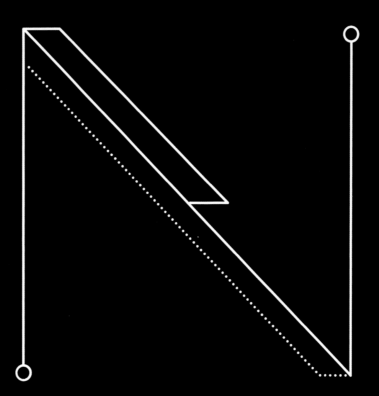

IS FOR

NASA

Not to be confused with the National Association for Stock Car Auto Racing, NASA describes its purpose as 'responsible for unique scientific and technological achievements in human spaceflight, aeronautics, space science, and space applications that have had widespread impacts on our nation and the world'.[1] There's little doubt that the last sixty years has seen NASA undertake numerous feats of dazzling technological achievement. However, I find it somewhat ironic that an institution so deeply rooted in some of humanity's best qualities – the desire for exploration, knowledge and *looking out* into the Universe – was, in fact, born of some of our worst: paranoia, conflict and the threat of war.

On 4 October 1957, the Soviet Union launched the first human-made satellite, named Sputnik 1, into Earth's orbit. Despite its short lifespan of only three weeks, its impact on the political stage was enormous, and symbolic of far more than just our desire as a species to explore space. The fact that the first satellite to enter Earth's orbit was launched by the Soviet Union was completely at odds with the United States' self-perception as the world's most technologically superior nation. While the US government was well aware of the Soviet Union's space-flight capabilities prior to the launch of the satellite, the public were naturally less informed, so the Soviets' progress came as a shock. Sputnik 1, a symbol of the threat of communism, represented a decisive victory in the first leg of the Space Race; the battle for dominance between the United States and the Soviet Union in space being just one part of the wider Cold War between the two nations.[2]

Quick to show the world that the United States may not have been the first to enter a satellite into orbit but were still very much in the race, just a year later, in 1958, the United States government announced the formation of the militarily focused Advanced Research Projects Agency (renamed DARPA in 1972) and the civilian space programme, the National Aeronautics and Space Administration.[3]

Just a few months into his presidency in 1961, John F. Kennedy made his ambitions for the United States space programme clear in a speech to Congress, stating it was 'time for this nation to take a clearly leading role in space achievement, which in many ways may hold the key to our future on Earth. I believe that this nation should commit itself to achieving the goal before this decade is out of landing a man on the Moon and returning him safely to the Earth. No single space project in this period will be more impressive to mankind or more important to the long-range exploration of space.'[4] Such lofty ambitions were certainly not political rhetoric. On 20 July 1969, shortly before the decade was complete, Neil Armstrong became the first person to set foot on the Moon – the first person to set foot on another world. It's easy to underestimate how far away the Moon is and just what an incredible feat it was to not just propel people off the planet and then land them on the Moon, but to cover such a huge distance. Even though the Moon is our closest neighbour, it's still over a 380,000 km away – the equivalent of flying around the world nearly ten times. What's even more impressive is that this was achieved with technology that was incredibly basic, even compared to the phone you have in your pocket. A standard iPhone has over 100,000 times the processing power of the Apollo Guidance Computer aboard the Apollo 11 spacecraft Eagle,[5] the module that landed on the Moon.

The Apollo 11 mission was watched by millions around the globe. Robert Wussler, CBS's executive producer of the mission's television coverage, declared it would be 'the world's greatest single broadcast'.[6] Despite the Soviet Union's own significant achievements, including landing the first human-made object on the Moon, taking the first images of the far side of the Moon in 1959 and then sending the first man into space in 1961, it was NASA's Apollo 11 mission that marked the climax of the Space Race. Including Apollo 11's iconic mission, NASA has successfully orchestrated six Moon landings between 1969 and 1972 with a total of twelve men having walked on the Moon's surface.

While NASA's first decade was largely defined by the moonwalk, as I like to call it, that is only one (albeit colossally significant) part of its contribution towards our exploration of the Universe. While the space agency's achievements are too numerous to list in this modest digest, there are some missions of particular significance for a space tourist, such as I, to note.

The Hubble Telescope, which launched in co-operation with the European Space Agency in 1990, has delivered over a million observations, including discovering new moons of Pluto, helping to pinpoint the age of the Universe, confirming that the rate of the Universe's expansion is accelerating, and providing evidence that most galaxies contain black holes at their centre. The pioneering probe *New Horizons* was launched in 2006, conducting a fly-by study of Pluto and its moons, and is now travelling deep into the Kuiper Belt, almost 8 billion km from Earth – further than any human-made object has gone before.[7] NASA's Mars Exploration Program has seen extensive analysis of the Red Planet with orbiting satellites and rovers on the planet's surface. The programme has previously investigated a multitude of areas, including weather patterns, the planet's surface, and chemical analysis. Present missions include the recently landed rover *Perseverance*, which aims to answer whether Mars could ever have supported microbial life.[8]

But what is next for NASA? The emergence of companies offering privately funded space travel, such as SpaceX, means that NASA will be increasingly collaborating with commercial companies to fulfil its goals of returning to the Moon, placing human feet on Mars, expanding our knowledge of the solar system and beyond, and potentially even answering the question, 'Are we alone in the Universe?'

'You develop an instant global consciousness, a people orientation, an intense dissatisfaction with the state of the world, and a compulsion to do something about it. From out there on the moon, international politics look so petty. You want to grab a politician by the scruff of the neck and drag him a quarter of a million miles out and say, "Look at that."'

EDGAR D. MITCHELL

LUNAR MODULE PILOT OF APOLLO 14

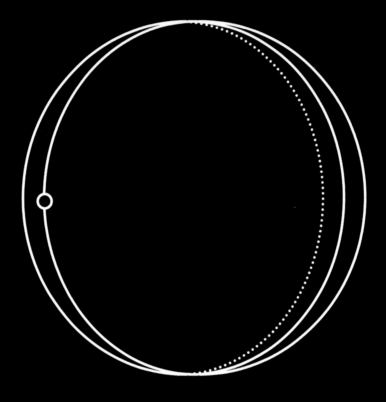

IS FOR

Overview Effect

We've all felt small shifts in perception at times; the spiritual calm that washes over you while witnessing a particularly beautiful sunset, the feeling of insignificance when looking up at the stars on a clear evening, or even returning from a holiday feeling passionate about quitting your office job immediately to become a professional scuba instructor. For the most part, however, these feelings are temporary. But imagine, for a moment, a shift in perception so significant that it left you fundamentally changed for ever.

Ptolemy thought we were very literally at the centre of the Universe, that our humble planet served as the epicentre for all that exists – a pivot for all that there ever was, is and will be. The Universe rotating around us through eternity. We now know this to be untrue – that we are but a splash of blue against the cosmic canvas – but that doesn't prevent us as individuals from feeling psychologically at the centre of our own universe, burdened by the daily routine of essential tasks and procedures that define our day-to-day life. Beholden to our egocentric perception of the world through necessity, we largely experience a limited, myopic view of our species and our place in the Universe. After all, we all have lives to live, jobs to keep, bills to pay, relationships to maintain. It's difficult – perhaps impossible – to benefit from a truly macro view of our existence. David Beaver, co-founder of the Overview Institute, a community dedicated to the shift in global consciousness experienced through space flight, described the sentiments of one of the Apollo 8 astronauts who were the first to orbit the Moon: 'When we originally went to the Moon, our total focus was on the Moon. We weren't thinking about looking back at the Earth. But now that we've done it, that may well have been the most important reason we went.'[1]

The phrase 'overview effect' was coined by space philosopher Frank White, and refers to a phenomenon experienced by astronauts who have a profound shift in their perception of Planet Earth and our species' existence after seeing our planet from space. The concept of the overview effect occurred to Frank White when he was flying across the US and he began to imagine life outside our atmosphere. He imagined looking back down on Planet Earth, the ability to see Earth as a whole, the ability to gain a holistic perspective of the planet: 'I would always have an overview of the Earth. I would see it from a distance. And I would see it's a unified whole. There are no borders or boundaries. All of these things would become knowledge. Which, living on the surface, we find it very hard to philosophically grasp, or mentally grasp.'[2] Of course, we are cognitively capable of understanding that the borders between countries are social constructs and that humans are very much one part of the planet's bio-system, not separate from it. But understanding the facts is different from feeling them on a deeper level, which is what those individuals who have left the planet have described.

Colonel Chris Hadfield, who spent 166 days on the International Space Station (ISS), described his experience of circling the Earth more than 2,600 times:

> *Every circumnavigation is like another voyage of discovery. And the angle between you and the Sun and the world is changing ... Each time around, you see the world slightly more for what it is ... and somewhere along the way, in one of the thousands of orbits, I think you fundamentally – I don't know if it's shifted or deepened, or both, but you start to see the world for what it is.*[3]

Witnessing the seasons change within a day, watching thunder and lightning storms flicker in the clouds, seeing your own home country as just one small part of the patchwork of the global landscape – all leave an indelible mark upon those who are able to experience them.

It's safe to say that many of the planet's problems could be solved if more people were able to experience the overview effect and then carry its virtues with them. However, with the overview effect being ostensibly only achievable by physically leaving the planet, will it ever be experienced to some degree by enough people to make a difference here on Earth? Space tourism will only be available to the super-rich for the foreseeable future: Jeff Bezos's Blue Origin is commanding $55 million for a trip to the ISS.[4] Given this, perhaps our most viable option is to utilise technology to help more people undergo a shift in perception of their own.

Immersive technologies involving virtual reality are providing clear avenues for getting as close to space as possible while still remaining on Planet Earth. The multi award-winning immersive VR experience HOME: A VR Spacewalk, created by the BBC and REWIND, is 'as close as it gets to being in space', says astronaut Nicole Stott.[5] This technology allows users to undertake a fifteen-minute spacewalk on the ISS via an Oculus headset. Another option for experiencing space involves a twist to a current relaxation trend. For me, the name 'isolation tank' sounds more akin to an interrogation technique than recreation. The idea of being left alone with one's thoughts in the darkness, floating in salty water for an hour, is not my cup of tea (or recycled coffee, if you've read 'I is for International Space Station'). However, a company called SpaceVR, for which Frank White is an advisor, has combined the popular concept of flotation tanks and combined it with virtual reality. The process allows people to experience the feeling of zero gravity by floating in a saline-solution-filled capsule while simultaneously watching footage from the ISS via a VR headset. SpaceVR are even planning on launching their own satellite, Overview 1, to provide their own footage from low orbit into the experience.[6]

Edgar Mitchell, lunar module pilot for Apollo 14, said,

> *"You develop an instant global consciousness, a people orientation, an intense dissatisfaction with the state of the world, and a compulsion to do something about it. From out there on the moon, international politics look so petty. You want to grab a politician by the scruff of the neck and drag him a quarter of a million miles out and say, "Look at that."'* [7]

With something as profound as the overview effect, which appears to resonate so deeply with those who have experienced it as to alter their entire world view, who knows what we might achieve and how differently we might treat each other, treat our planet, and change the course of our existence if we could all share even some of this cosmic perspective.

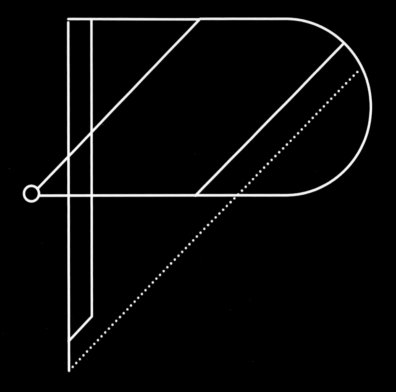

IS FOR

Pluto

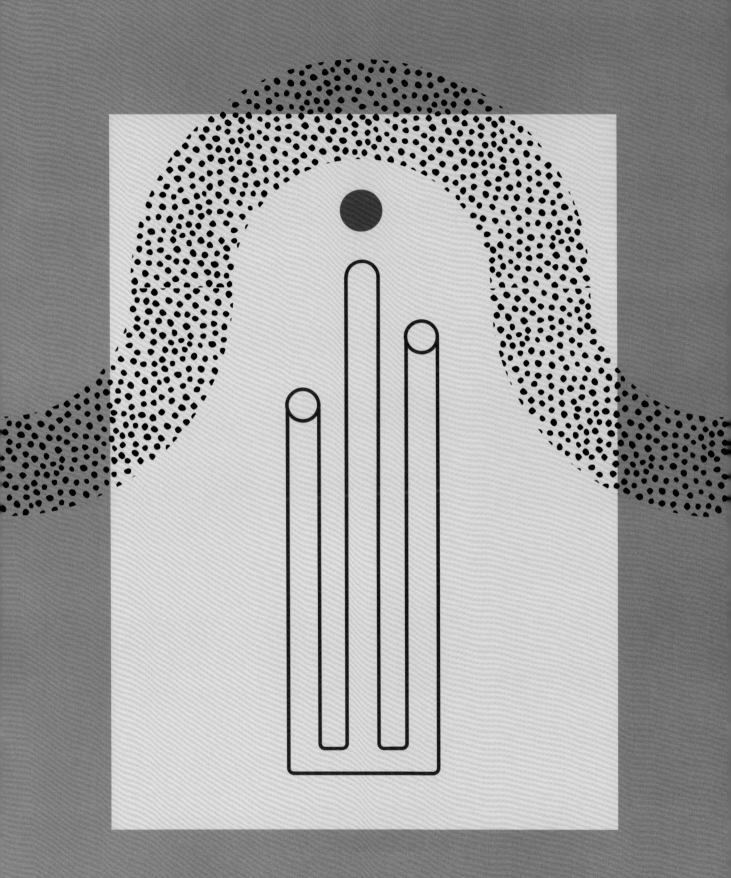

For a relatively small, rocky sphere on the outskirts of the solar system, Pluto has managed to generate more contentious headlines than any other single body in our galaxy, with NASA bosses, angry schoolchildren, world-famous astrophysicists and even legendary rock stars all having their say. Is it all just a storm in a teacup? Not if you're Pluto. How would you like to have your physical attributes discussed globally and then be unceremoniously demoted in such a public fashion? 'Why is Pluto No Longer a Planet?'[1] asked the BBC in 2015. 'Should Pluto Be a Planet Again?'[2] Forbes asked in 2019. The debate as to Pluto's planetary status has been one that has lasted for some time.

As with many cosmological discoveries, Pluto's existence was first suggested through its effects on other bodies, rather than being observed directly itself. In the early 1900s, American astronomer Percival Lowell saw unusual deviations in the orbits of Neptune and Uranus and thought that these orbital irregularities were caused by the gravitational influence of a ninth planet. Lowell searched for many years to identify this ninth planet but was unsuccessful.[3] Where mere mortals may begin shouting obscenities after ten minutes searching for their lost car keys, astronomy requires a level of stoic patience that most of us can only dream of. It wasn't until 1930 that Clyde Tombaugh discovered Pluto at the very observatory in Arizona that carried Percival Lowell's name.[4] The observation of Pluto was significant for several reasons: not only was it the ninth planet to be found in the solar system, but it was also the first Kuiper Belt object to be discovered – although the astronomers at the Lowell Observatory didn't realise that at the time. It wasn't until sixty-two years later that the second Kuiper Belt object was discovered, confirming that

Pluto was one of many objects circling the Sun in the outer regions of our solar system.[5] The name 'Pluto' was given to the (former) ninth planet by an eleven-year-old schoolgirl called Venetia Burney from Oxford. She thought that the Roman name for the god of the underworld was a befitting moniker for this distant, icy planet.[6]

In 2006, seventy-six years after being named the solar system's ninth planet, Pluto had its planetary status revoked and was given the new classification 'dwarf planet'. The International Astronomical Union (IAU) stated that Pluto did not meet the full criteria necessary to achieve full planetary status. According to the IAU, a planet must meet three criteria to be deemed a planet – it must orbit a star, be of sufficient gravity to force itself into a spherical shape, and of enough size to have done away with any other objects of a similar size in or around its orbit.[7] Pluto, while meeting the first two, fell at the final hurdle and hasn't managed to clear its orbit.

The 2006 downgrade wasn't the first time Pluto's status as a planet had hit the headlines, with the *New York Times* running the headline in 2001, 'Pluto's Not a Planet? Only in New York.'[8] The *New York Times* was reporting on the Hayden Planetarium's decision to not include Pluto in its list of the solar system's planets and instead group Pluto with other Kuiper Belt objects. Neil deGrasse Tyson, director of the Hayden Planetarium, wrote an open letter pointing out that Pluto's demotion was not a demotion at all, just a reclassification based on the evolving definition of a planet, giving a better representation of the structure of the solar system. He also argued that Pluto had not been 'kicked out' of the solar system at all:

I have previously written on the subject, in an essay titled 'Pluto's Honour',[9] where I review how the classification of 'planet' in our solar system has changed many times, most notably with the 1801 discovery of the first of many new planets in orbit between Mars and Jupiter. These new planets ... later became known as asteroids. In the essay ... I argued strongly that Pluto, being half ice by volume, should assume its rightful status as the King of the Kuiper Belt of comets.[10]

Such rational arguments, however, didn't prevent deGrasse Tyson from receiving many letters from irate schoolchildren. One particularly angry letter said, 'I think Pluto is a planet. Why do you think Pluto is no longer a planet? I do not like your answer!!! Pluto is my faveret [sic] planet!!! You are going to have to take all of the books away and change them. Pluto is a planet!!!!!!! Your friend, Emerson York.'[11]

Pluto once again piqued the public's interest with the launch of the *New Horizon* spacecraft, NASA's first mission to study Pluto, its moons and the Kuiper Belt. Leaving Earth in 2006, *New Horizons* took nine years to reach its target, and the results of its six-month study were outstanding. They include the discovery that Pluto's moons are all the same age, implying they were all created by a single collision with another Kuiper Belt object;[12] that a glacier on Pluto discovered by *New Horizons* is the biggest in the solar system; and that Pluto may well have an internal water-ice ocean. Overall, Pluto and its moon are far more complex than previously imagined.[13] Following its reconnaissance mission on Pluto, *New Horizons* was far from done and is set to discover more of the icy, dark region of the solar system, travelling deeper into the Kuiper Belt more than 6 billion km from Earth.

The *New Horizons* mission has undoubtedly helped continue Pluto's planetary parley in recent years, with a NASA administrator jokingly stating that Pluto is a planet 'once again',[14] and *New Horizons* lead, Alan Stern, debating the ex-president of the IAC about both Pluto's status and the very definition of what constitutes a planet, Stern arguing that our current definition is fundamentally flawed.[15] Even Brian May, the guitarist from Queen, who received his doctorate in astrophysics, threw in for the icy orb, saying in an Instagram post, 'Let's hear it for PLUTO – the 9th Planet!!'[16]

Whether it be the preschoolers' favourite planet or King of the Kuiper Belt, Pluto has proved to be a mysterious and fascinating object following its discovery and has caused debate ever since. Regardless of how Pluto's classification evolves in the coming decades, any subject that inspires children to write impassioned letters to world-famous astrophysicists can only be a cause for celebration.

'Although quantum mechanics is an extraordinarily successful scientific theory, on which much of our modern, tech-obsessed lifestyles depend, it is also completely mad.'

JIM BAGGOTT

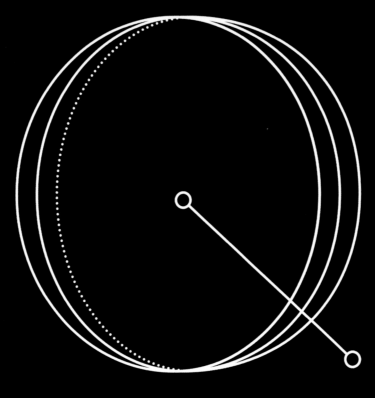

IS FOR

Quantum Physics

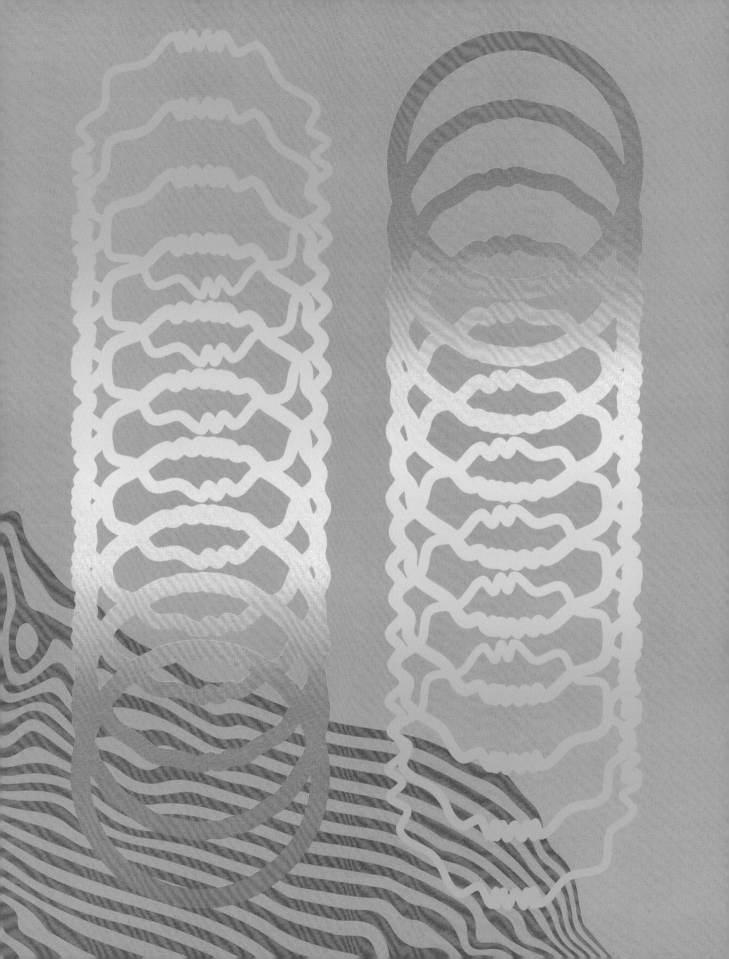

Throughout this book we've dealt with some extraordinarily big numbers and cosmic structures of incomprehensible proportions, but this chapter is not only the complete opposite (focusing on the phenomenally small), it's also potentially the most intimidating subject to get one's head around. Even physicists would admit that when entering the quantum realm, you encounter behaviours that are as mysterious and confounding as they are fascinating. However, quantum physics is not an abstraction, something to be pondered from afar; quantum physics is a part of everyday life, responsible for the laws that govern our universe at the atomic and subatomic level... and the very reason we exist – even if the most popular word associated with quantum is *spooky* (thanks to Albert Einstein). Why spooky? Well, quantum physics is not only a branch of physics that is largely incompatible with classical physics and therefore remarkable in itself, but there are manifestations of the quantum world which, to put it simply, are astounding.

With such an impactful branch of physics, one which examines the very fabric of the Universe, it's surprising to know that its discovery was largely associated with the common light bulb. While several brilliant minds contributed to the development and understanding of quantum physics, it is the German theoretical physicist Max Planck who is today credited with its discovery, even though, at first, he admitted that he did not understand it.[1] Planck's discovery in 1900 originated from his experiments surrounding light-bulb efficiency, which led him to realise that light waves carry energy in 'quanta', a word derived from the Latin for 'packets', and that energy can show characteristics of matter.[2] It appeared that Planck was largely unaware of the broader implications of his discovery and how it would go on to revolutionise our understanding of physics entirely. Despite Planck being attributed with discovering quantum physics

through his work, science historian Professor Helge Kragh said, 'the "discovery" should be seen as an extended process and not as a moment of insight communicated on a particular day in late 1900'.[3] Many scientists were involved in the development of quantum theory, including but by no means limited to Albert Einstein, Paul Dirac, Werner Heisenberg, Niels Bohr, Richard Feynman and Erwin Schrödinger. Each one is worthy of a far more in-depth reflection than this book allows for.

Richard Feynman, the Nobel Prize-winning astrophysicist, said, 'I think I can safely say nobody understands quantum mechanics.'[4] While true in some respects, this could be slightly misleading, as many applications of quantum mechanics are not only understood in precise detail but implemented on a daily basis. Smartphones, nuclear power stations, computers, lasers, digital cameras, alarm clocks, GPS, MRI scans[5] to name a few, all rely upon our comprehension of quantum physics. But Richard Feynman was playfully alluding to the deeper understanding of quantum mechanics, not its useful applications.

So, if we understand quantum mechanics in such detail that it has become an entirely indispensable part of our daily routines, why is it still a theory that attracts so much speculation, disagreement and mystery? The theoretical physicist Sean Carroll wrote for the *New York Times*, 'Scientists can use quantum mechanics with perfect confidence ... What we don't do is claim to understand quantum mechanics. Physicists don't understand their own theory any better than a typical smartphone user understands what's going on inside the device.'[6] There are several well-known conundrums that still have scientists scratching their heads when it comes to explaining exactly how or why quantum mechanics works in the way it does.

Superposition, for example, explains how a particle can exist in two entirely different states at once and then, once measured or observed, *collapses* into one of those states, implying that the act of measurement itself has influenced the outcome. This paradox was made famous by a thought experiment from physicist Erwin Schrödinger known as 'Schrödinger's Cat'. This experiment described a cat in a box, along with a device that had a 50 per cent chance of killing it. According to quantum mechanics, Schrödinger argued that just before the box is opened, the cat is both alive and dead simultaneously – inhabiting the aforementioned superposition. We, as observers, then force the cat to be either alive or dead when we open the box. Now, if this sounds slightly outlandish, that was Schrödinger's intention: to use a non-quantum metaphor for the oddity of the quantum realm.[7]

Quantum entanglement, or as Einstein described it, 'spooky action at a distance',[8] is the apparent ability of one object to instantaneously influence another object, even over great distances. A team of international physicists 'entangled' pairs of photons, separated the pairs by 143 km and then altered the state of one to see how quickly it would influence the state of the other. The results were incredible, showing that entanglement transports information at least four times quicker than the speed of light. In fact, the scientists did not rule out the possibility that the change in state of the entangled photons was instantaneous – their equipment just couldn't get a more accurate reading.[9] This poses a serious problem for Einstein's widely accepted theory that nothing can travel faster than light because the faster something moves, the more mass it acquires and the more energy it needs to power its acceleration.

Another quantum phenomenon, evocatively named 'quantum tunnelling', describes quantum particles' ability to 'tunnel' through barriers that they quite simply shouldn't be able to. The rules of classical physics would prevent the particle from moving through these barriers, but due to particles also behaving like waves at a quantum level, there is a small probability of the particle appearing on the other side. Quantum tunnelling is the very reason we have sunlight and therefore the very reason we are able to survive.

Our Sun works using nuclear fusion. Despite being a huge fiery ball of gas, it's not actually hot enough to give the photons sufficient energy to overcome their natural repulsion from one another for fusion to occur... the vital process for sunlight. Here's where quantum tunnelling and the laws of probability upon which quantum mechanics is based very much help us out. The YouTube channel MinutePhysics explains, 'Hydrogen atoms, even if they aren't quite hot enough for fusion, will fuse together anyway. And the Sun is so big and has so much hydrogen that these small chances happen all the time.'[10]

So, with phenomena such as quantum entanglement undermining Einstein's theories, it begs the question, how is this possible and what are we missing? This is where a hypothetical particle known as a 'tachyon' has been discussed. Tachyons, taken from the Greek word for 'swift', theoretically work within Einstein's rules by coming into existence already travelling faster than the speed of light, rather than speeding up and passing it.[11] However, there is no observational evidence as yet for the existence of the tachyon, and so questions surrounding some of the stranger elements of quantum mechanics remain unanswered.

I'll leave you with my favourite summary of quantum physics, belonging to science writer Jim Baggott, 'Although quantum mechanics is an extraordinarily successful scientific theory, on which much of our modern, tech-obsessed lifestyles depend, it is also completely mad.'[12] Amen to that, Jim. Amen to that.

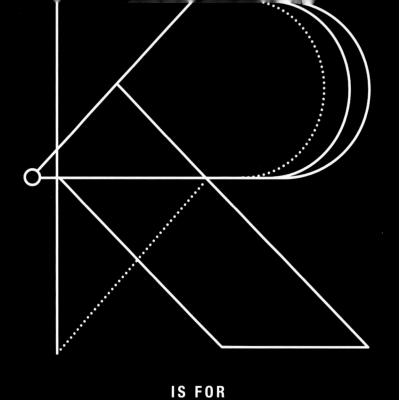

IS FOR

Red Planet

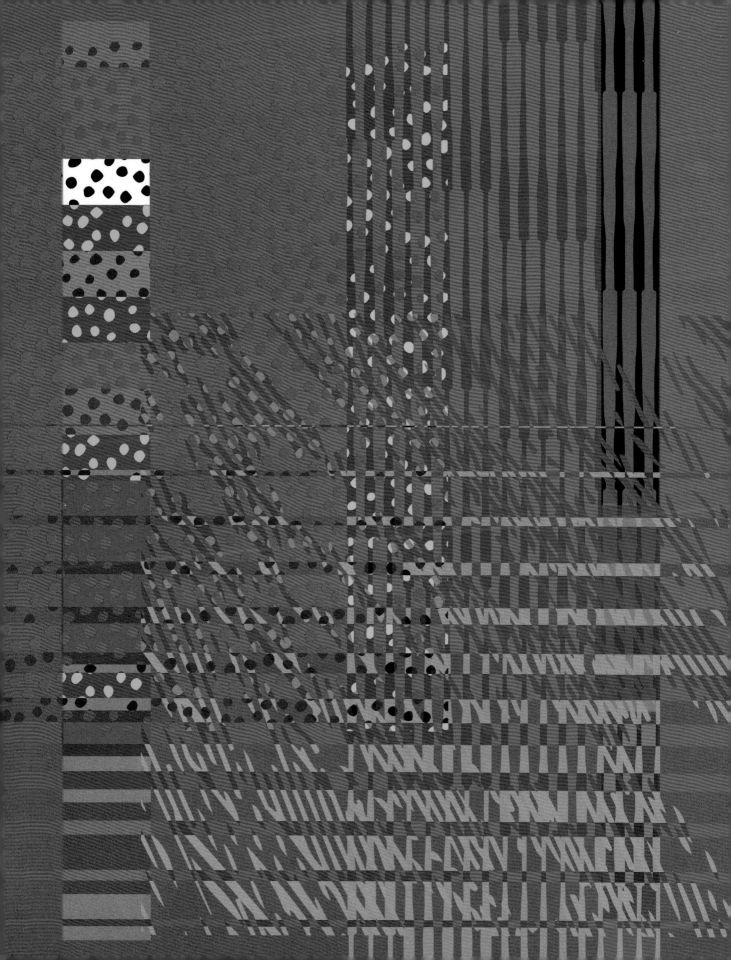

Mars Attacks! saw skeletal aliens conduct a global invasion; *Ghosts of Mars* witnessed Ice Cube and Jason Statham battle ancient body-snatching warriors; and when H. G. Wells' *War of the Worlds* was broadcast on BBC radio in 1938, people actually believed we were under attack by large machines from the heavens. The various depictions of Mars in popular culture have not always been the kindest – but then what can you expect with a planet who was named after the Roman god of war?

Like a blood-red eye staring down from the black, Mars has always been a point of fascination for us on Earth. Visible to the naked eye, and our closest planetary neighbour, the Red Planet may hold vital secrets of whether there's life in the solar system, how life started here on Earth and how life might continue if we were to ever leave. This is what makes Mars the most studied planet in the solar system, apart from Earth, with huge resources and effort having gone into understanding its nature.[1] Despite appearing to be a dead, dusty planet inhospitable to life, Mars' scarred surface tells a different story; a story that it was once host to an abundance of flowing water and perhaps even life.

The detailed knowledge we have today is owed to the incredible technical prowess that has enabled us to get up close and personal with the Red Planet through observations from orbital modules and rovers down on the surface itself. Between 1960 and 1964, a number of USSR missions and one from the United States attempted to fly spacecraft to Mars, but all failed. It was NASA's *Mariner 4* that successfully achieved a fly-by, returning twenty-one images of the planet's surface and paving the way for future missions to Mars. An observation that would become vital for future missions to Mars was that the planet had very low atmospheric pressure, meaning that to land anything on the surface would require retro-boosters and a parachute to reduce any approaching craft's speed.[2] Since 1965 the information gathered from spacecraft orbiting Mars has been extensive, studying weather patterns and climate change, documenting the surface with detailed photography and even searching for gases associated with biological activity. However, there are limitations to what data the orbiting crafts can capture, which is where the rovers step in... or should I say, fall, glide, then roll in.

Imagine, if you will, the difficulties and complexities of placing a fully functioning, autonomously operating mobile laboratory on a different planet millions of miles away. What it would take to design such a tool – a tool that can withstand leaving our atmosphere, flying through the solar system for several months, entering Mars' scorching atmosphere and landing unscathed on the surface... only then to begin the actual hard work.

The latest rover to land on the Martian surface is *Perseverance* (2021), whose mission is to search for ancient life and collect rock and soil samples that could, at some point, be returned to Earth.[3] At around 3 m long and weighing in at 1 tonne (a little more than 1979 VW Beetle), NASA needed to take great care in delivering this precious piece of technology to the surface in one piece. The craft containing *Perseverance* entered the Martian atmosphere at such speeds it required the strongest parachute ever built to drastically reduce its velocity, retro-boosters to slow it down even further and then what is known as a sky crane to lower the rover gently to the ground to begin its mission.[4]

Hitching a ride on *Perseverance*'s belly was a 2-kilo helicopter called *Ingenuity*. Shortly after landing, Ingenuity took flight, becoming, according to

NASA, 'the very first powered, controlled flight in the extremely thin atmosphere of Mars, and, in fact, the first such flight in any world beyond Earth'.[5] *Ingenuity* is providing vital information for future missions on how to fly in the Martian atmosphere and as of December 2021, already had eighteen flights under its belt.

So, what have we learnt about Mars so far and how is that benefiting our broader knowledge of the potential life off-Earth? Despite Mars' surface being lots of different colours, it appears red due to the rusting iron minerals in the soil. While Mars now seems to be a desolate, inhospitable world, the evidence shows that it was once a warmer, wetter place. There's a great deal of evidence to support this, including rocks and minerals on the surface that can only form in liquid water, river valley networks and lake beds.[6] Estimations based on observations from the creatively named Very Large Telescope in Chile even hypothesised that Mars was once host to an ocean that may have covered almost half of the planet's northern hemisphere – larger than Earth's Arctic Ocean.[7] Data from the *Curiosity* rover, which landed on Mars in 2012, also revealed that a water lake within Mars' Gale Crater held all the necessary building blocks for life.[8] There is still water on Mars today in the form of ice under the planet's polar regions, but water cannot exist on the surface due to its thin atmosphere.

But could Mars ever have supported life? While the question remains unanswered, there is certainly compelling evidence to suggest that life was once possible. In 2018 the *Curiosity* rover found organic material in 3-billion-year-old rocks that are not unequivocal evidence for life in their own right but are indicators that life may have once been existed. Thomas Zurbuchen, NASA's associate administrator for the Science Mission Directorate, said, 'With these new findings, Mars is telling us to stay the course and keep searching for evidence of life ... I'm confident that our ongoing and planned missions will unlock even more breathtaking discoveries on the Red Planet.'[9] The *Perseverance* rover, with the most technologically advanced on-board equipment seen on a rover thus far, will continue its predecessors' work, giving us the best opportunities for discovering any evidence of life. However, a key objective of *Perseverance* includes demonstrating technology that may be used for future human exploration. Given that Mars is extremely unwelcoming for human life, preparing the landscape for human explorers will be incredibly complicated, and so an intricate understanding of the Martian environment is essential for any successful visit involving people. Rovers like *Perseverance* and *Curiosity*, orbital spacecrafts including *Mars Odyssey* and MAVEN, and landers such as *InSight,* all utilise their plethora of intricate tools to paint the detailed picture necessary for a deep understanding of this once mysterious planet.

'2021 was an epic year for space exploration,' said Space.com's Mike Wall,[10] referring to a series of major international missions to Mars. In addition to NASA's *Perseverance*, Mars also welcomed the United Arab Emirates' Mars Mission, which aims to build a complete picture of the Martian atmosphere. China's first Mars Mission also launched in 2021. *Tianwen-1* (meaning 'questioning the heavens') comprised of six spacecraft, including a rover called *Zhurong* (God of War). The mission will investigate the Martian soil, the planet's geology, its atmosphere and water-ice distribution.[11]

Despite the frequent depictions in popular culture of Martian marauders poised to colonise our planet, our clear intention to send humans to Mars would imply that the invaders are most likely to be us.

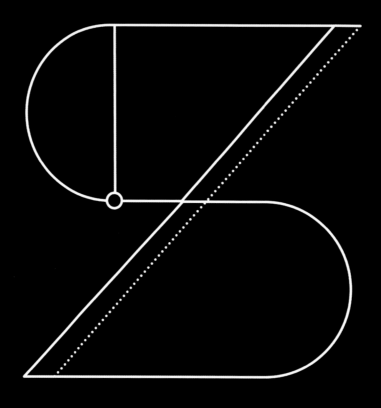

IS FOR

SETI Institute

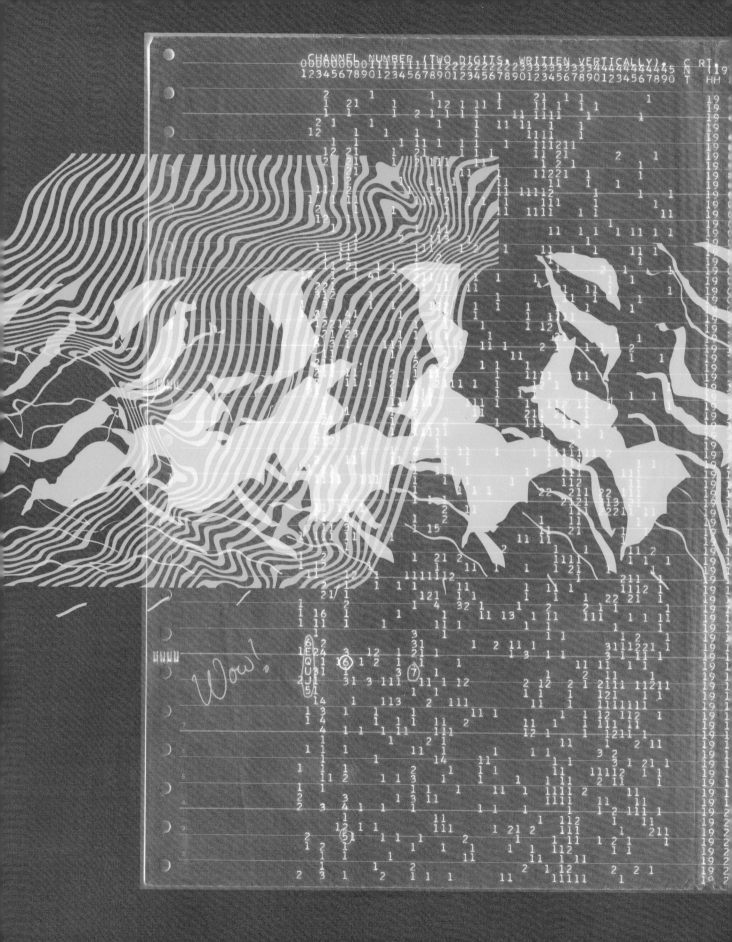

If you are looking for a career that provides instant and frequent gratification, joining the search for extraterrestrial intelligence probably isn't for you. However, the discovery of life elsewhere in the Universe, intelligent or otherwise, would undoubtedly be the most extraordinary and profound discovery of all time – ample reward for such a difficult task where the odds of discovery, at least in the short term, are ostensibly stacked against us.

In 1961 an astronomer called Frank Drake, who has since been described as 'the father of SETI science',[1] led the first attempt to detect interstellar intelligent communication by directing a radio telescope at the Cetus and Eridanus constellations.[2] Drake and his team were looking for uniform pulses that would indicate this. Unfortunately, none were detected but this search marked the first mission to pave the way for the ongoing search that has continued since.

While the hunt for extraterrestrial intelligence has been formally underway since the 1960s, the most prolific organisation in this field is known as the SETI (Search for Extraterrestrial Intelligence) Institute. Formed in 1984, the SETI Institute's mission is to examine the origins of life in the Universe and how intelligent life may develop. It seeks to do this through finding beings that are, at the very least, as technologically advanced as we are, utilising scientific fields such as astrobiology and methods including searching for signals from other civilisations.[3]

The search for extraterrestrial intelligence is often misunderstood and regularly conflated with UFOs and sensationalist headlines from mainstream media. Jason T. Wright, a professor of astronomy and astrophysics at Penn State University, commented in a blog post, 'It's pretty embarrassing to see your work so brazenly sensationalized in the media ... I've developed a thick skin about it, but it still smarts to see my name next to pictures of bug-eyed aliens.'[4] Talk of abductions, little green men and crop circles are a reductive and misleading representation of the SETI Institute's work, and only serve to fortify perception of the institute in some circles as a fringe science. According to NASA policy analyst Stephen J. Garber, one of the primary reasons NASA withdrew its funding from the SETI Institute was because 'the "giggle factor" had persisted in making SETI a perennially easy political target. While SETI involved truly fundamental science, it didn't fit neatly into any scientific discipline.'[5]

When searching for intelligent life, there are multiple different indicators that SETI takes into account. The origins of the search for extraterrestrial intelligence involve listening for direct communication through radio signals, and have evolved to far systems, such as those deployed by the Allen Telescope Array (ATA) in California. The ATA was designed specifically for the SETI Institute's purposes and, thanks to a number of donors (including the high-profile donor for whom it was named, Microsoft co-founder Paul Allen), this telescope can operate the institute's research full- time and over multiple areas of the Universe at any one time.[6] But this is not the only method employed; another practice is the search for biosignatures – an instance of which gained some media attention in 2020 when phosphine gas was reportedly detected in the atmosphere of Venus. As phosphates are only naturally created on Earth via microbes that live in oxygen-free environments, this was an intriguing biosignature that implied the possibility of life on the second planet from the Sun. However, doubts have since been cast on this observation and the phosphine is more likely to be sulphur dioxide.[7] This links to the SETI Institute's research in astrobiology, studying the origins and development of life, and organisms that can survive extreme conditions here on Earth, to gain a better understanding of how life might form on other planets.

SETI scientists also look for megastructures indicative of an intelligent civilisation, which some argue may be a more fruitful method of searching for evidence of intelligent life than waiting for a moment of synchronicity where we happen to receive a signal at the exact moment and from the exact direction we are looking in.[8]

It is this battle with probability and reliance on some form of synchronicity that makes the search for life in the cosmos, let alone intelligent extraterrestrial life, so very, very difficult. But scientists associated with the search for extraterrestrial intelligence are confident in the future of the search. Professor Michael Garrett, vice chairman of the International Academy of Astronautics and SETI Permanent Commission, states that the technology associated with radio astronomy 'improves exponentially. Every couple of years our capacity to detect a SETI signal doubles.'[9] It's these rapid improvements to telescope capabilities and other emerging technology, such as using artificial intelligence to monitor incoming signals, that will greatly improve our chances of detecting signals of significance.

While there's been no evidence of intelligent alien life so far, there have been a number of interesting signals since the search for life in the cosmos began, perhaps none more famous than the 'Wow!' signal of 1977. Due to the unusual and seemingly inexplicable nature of the seventy-two-second signal, the radio astronomer who discovered it, Jerry Ehman, wrote 'Wow!' next to it. The signal remained a mystery until recent years, when an astronomer wrote a paper attributing the signal to a comet that had not been discovered in 1977 – although this theory has been challenged by numerous astronomers, including Ehman himself.[10]

But what if the stars were to align both literally and metaphorically and we were to receive a signal that bore all the characteristics of an intelligent civilisation? The SETI Institute has a set of protocols in place should this ever happen. The evidence must first be verified, ideally through observations at different facilities, and once the evidence has been corroborated, the information is to be released to the international scientific community in a transparent manner. Does this mean anyone can don their 'Welcome to Earth' t-shirt, rush to their garden shed and start transmitting their thesis on intergalactic relations into the cosmos by way of response? In short, no. The confirmed signal would be monitored by an International Academy of Astronautics and SETI task group to offer ongoing support and advice, but there would be no response to the signal without first consulting an international body such as the United Nations.[11]

With the exponential increases in technological developments aiding the search, SETI's future successes will be largely associated with continued funding. The Breakthrough Listen Initiative, funded by Yuri and Julia Milner, is the largest and most comprehensive programme to find evidence of alien civilisations, surveying 1 million of Earth's closest stars. This programme, which uses the world's most advanced tools, will cost $100 million.[12] To scratch the surface of the expanse of the Universe will require continued and significant private and public funding.

The search for intelligent life in the Universe continues to be a long and difficult process. However, with some estimates suggesting there might be as many as 6 billion Earth-like planets in the Milky Way,[13] chances are, we're not alone. Carl Sagan, the king of succinct cosmological communication, has often been quoted as saying, 'The Universe is a pretty big place. If it's just us, seems like an awful waste of space.'[14]

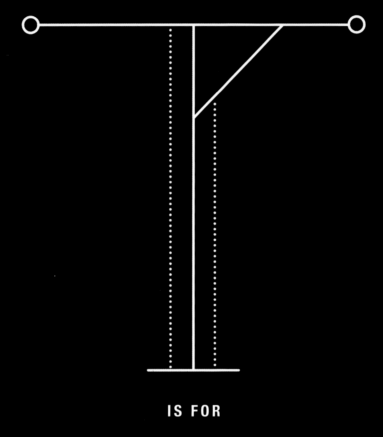

IS FOR

Trash

Unfortunately, where there's humans, there's trash. And space is no different. Junk, debris, trash, refuse, detritus, scrap, dross, litter: whatever you want to call it, there's a great deal of it in space – and we put it there.

Space trash is the name given to parts of satellites, rockets or any other type of human-made machinery that is no longer in use and has been left in space (nuts, bolts, paint flecks, etc.). While space is a very big place, a great deal of this space junk is currently orbiting at very high speed around our planet. According to the director of the Surrey Space Centre, Guglielmo Aglietti, there are around 8,000 tonnes orbiting Earth,[1] and while paint flecks don't sound all that intimidating, when they're travelling seven times faster than a speeding bullet[2] they can cause an incredible amount of damage.

Unfortunately, once in space, the human-made extraterrestrial tut doesn't just meander benignly through the cosmos. With an ever-increasing number of government and commercial satellites being launched into space, the chances of accidental collisions and explosions are only increasing, and the problem of space junk is going to get worse before it gets better. The Kessler effect, named after the NASA scientist Donald J. Kessler, who proposed the idea, describes the possibility of the exponential increase in space debris over time. Each collision creates more debris, and therefore increases the chance of more collisions and then more debris. This would result in a belt of debris around Earth, making any future endeavours in space far more difficult and dangerous.[3]

One of the largest space collisions took place in 2009, when an active communications satellite collided with a retired Russian military satellite, resulting in a whopping 1,800 pieces of debris joining the orbital dance around our planet. While near misses and minor collisions are frequent, in October of 2020 a much larger collision was worryingly close to taking place between part of a Chinese rocket and an inactive 1989 Russian navigation satellite. Daniel Ceperley, the CEO of LeoLabs, which provides data on low-Earth-orbit objects, said the collision would have caused clouds that would have 'spread out into a shell of debris around the Earth', leaving the trash there for centuries.[4]

Further debris has been caused by nations destroying their own satellites in vulgar displays of military power, publicly posturing with their anti-satellite technology – a rather expensive version of a gorilla beating its chest. But these demonstrations do more than damage international relations; these targeted explosions create thousands and thousands of smaller fragments of potentially dangerous junk. The Chinese anti-satellite explosion of 2007 is thought to have created more than 3,000 pieces of debris.[5]

But what are the immediate dangers of space debris to humans? As recently as September 2020, the International Space Station had to shift its orbit to avoid a piece of junk. And while there was no immediate threat to the astronauts' lives, those on board were still compelled to move to escape pods in case of a collision.[6] There's also the issue of space debris re-entering Earth's atmosphere and striking someone on the surface. Most of the debris entering the Earth's atmosphere burns up, but 20 per cent still makes its way through, and while the odds of being space-clobbered are slim (the only known incidents of people being hit by debris were in 1969 and 1997),[7] the amount of debris accumulating only increases the chances as time goes on.

The most tangible impact on the vast majority of people on Earth is our reliance on functional satellites and their vulnerability to collision with space junk. Communication, GPS, internet, television and weather forecasts all depend on satellites to function within our lives, and they are all reliant on not being obliterated by a piece of scrap metal travelling thirteen times faster than Concorde. The last thing we need is space junk causing people to smash their laptop into Earth junk in fits of rage because they can't download the latest episode of *Love Island* quickly enough.

But before you look to compose an angry tweet directed toward your local governmental representative, all is not lost. Scientists from across the globe are looking into practical solutions to remove debris from orbit so that space isn't destined to resemble the trash compactor from the Death Star just yet. They range from space nets, harpoons, robotic arms and even lasers to push debris back towards Earth to burn up in orbit. But all these solutions are complicated to undertake and extremely expensive, so any future solution will need all nations and companies involved with putting anything into space to take responsibility for taking it back out again safely. SpaceX's chief operating officer, Gwynne Shotwell, explained how the concerns around space debris has led SpaceX

to lower the altitude of their satellites so they will decay and burn up in Earth's atmosphere much more quickly. Shotwell also added that SpaceX's Starship could help remove space debris that's already circling the planet[8] like a benevolent rocket-powered litter-picker. And rightly so, given that private companies with commercial interests in space are likely to be major contributors to the future of cosmic congestion.

Next time you see someone flagrantly tossing an empty crisp packet to the ground and you feel that familiar pang of incandescent rage at the sheer anti-social audacity of it, take some solace that at least the litter isn't travelling at 29,000 kph.

'The cosmos is also within us. We're made of star-stuff. We are a way for the Universe to know itself.'

CARL SAGAN

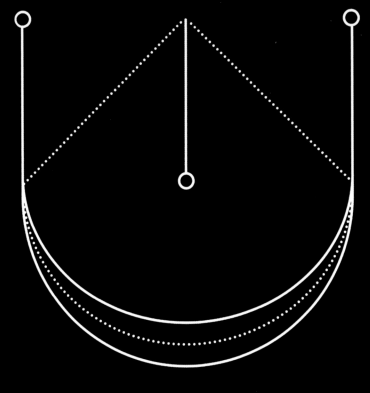

IS FOR

Universe

We might be part of a multiverse, our own realm of existence one of a trillion bubbles on a foamy sea of universes.[1] We might be in the centre of a black hole that resides in another, larger universe.[2] We might even be part of an elaborate simulation, the complicated construct of an omniscient architect's imagination.[3] However, based on what we know, it seems as if the Universe is everything. Everything there ever was, everything there is and everything there will be.

Without sounding too much like an elaborate riddle, if the Universe is everything, how long has everything been around for? Scientists use a number of methods to determine the age of the Universe, including observing its most ancient stars and clusters. Astronomer Dr Andrea Kunder told Space.com, 'Just like archaeologists use fossils to reconstruct the history of the Earth, astronomers use globular clusters to reconstruct the history of the galaxy.'[4] Scientists also look at the cosmic microwave background, which may sound like a 1960s psychedelic rock band but is actually the leftover radiation from the big bang – the huge explosion that is thought to have started the Universe's existence. *Explosion* is a loose term, though, as the Universe wasn't so much exploding as rapidly creating space in all directions.

From looking at the properties of the cosmic microwave background, scientists have been able to place the age of the Universe at around 13.8 billion years – a duration of time which most humans are pretty much incapable of comprehending. Cosmology is certainly not an ideal profession for meganumerophobics. (And yes, I did google that word to look clever.) If the whole existence of the Universe right up until now was represented in a twenty-four-hour period, humans only showed up at 23:58.[5] To date, our tenancy in this small part of the Universe has been rather short, to say the very least.

But what about the Universe's size? Well, as you may have guessed, it's big. And getting bigger all the time. Following observations of the cosmic microwave background, data from supernovae (ancient exploding stars), surveys of the patterns of galaxies and fluctuations in the density of the Universe, the estimated radius of the observable Universe is thought to be over 46 billion light years.[6] Unfortunately, the light from any part of the Universe beyond that will not have had the time necessary to reach us.[7] However, the evidence points to there being a huge amount of the Universe outside that radius which we are, as yet, unable to see. Astrophysicist Ethan Siegel wrote,

If we were to add up all the galaxies in the parts of the Universe that we'll someday see but cannot yet access today, we might be shocked to learn that there are more yet-to-be-revealed galaxies than there are galaxies in the visible Universe. There are an additional 2.7 trillion galaxies waiting to show us their light, on top of the 2 trillion we can already access.[8]

When contemplating what the Universe is actually made of, you might be surprised to learn that only a tiny amount constitutes the things we can see. The planets, stars, galaxies; these are all ordinary matter and make up only 4.9 per cent of the Universe. The remainder includes 26.8 per cent dark matter, which interacts with ordinary matter through gravity, and 68.3 per cent is made up of dark energy, which fuels the expansion of the Universe.[9] But as we've discussed earlier in this book, we don't really know what dark matter or dark energy is, just their unmistakable influence.

Following the work of scientists such as Vesto Slipher, Edwin Hubble and Georges Lemaître, we know that the Universe has been expanding ever since the big bang due to the light emitted from distant galaxies. When a galaxy is moving away from us, the light waves are stretched and appear red. The faster a galaxy moves away, the redder the light appears. So, if the Universe has been constantly expanding, then, according to traditional big bang theory, 13.8 billion years ago all the matter in the Universe would have been packed into a single point: the singularity – a point that burst into existence, creating space, time and all the matter in the known Universe, like a cosmic version of a Mary Poppins handbag.

While most of the fundamentals of the Universe were created in its early moments, millions of years passed before larger-scale cosmic formations arrived. It wasn't until around 200 million years after the big bang that the first stars began to form.[10] The stage for our own existence had to wait a great deal longer to be set, as it took 9.2 billion years for Earth to coalesce from the solar system's gas and dust.[11]

What does the future hold for the Universe? With dark energy fuelling its relentless outward march, it appears that the expansion will never cease, stretching the space between all the galaxies and stars further and further apart. Lawrence Krauss, astrophysicist and author, wrote for *Smithsonian Magazine*, 'The current expansion will continue forever, gaining speed, so that all the galaxies we now observe ... will one day disappear beyond our ability to detect them. Our galaxy will be alone in the visible universe. And then, once the stars burn out, the Universe will be truly cold, dark and empty.'[12]

Remember to look up and marvel at the unmitigated beauty of the night sky – the cosmic vistas to be witnessed by our distant-future descendants will shine less bright than the one we are fortunate enough to enjoy today, the stretched abyss of space too vast for light to overcome. Those who may surf the cosmos in millennia to come will sadly do so in darkness.

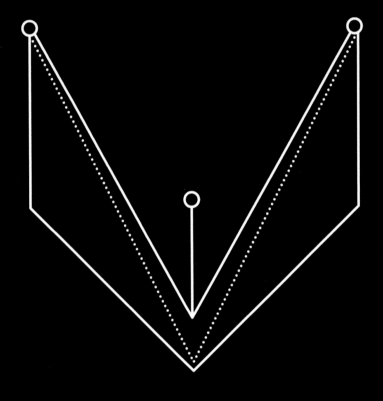

IS FOR

Voyager's
Golden Record

The Voyager mission was and still is one of the most remarkable space missions of all time for a number of reasons. The spacecraft *Voyager 1* is now the most distant human-made object ever created, currently travelling at over 56,000 kph into interstellar space. Along with its twin, *Voyager 2*, they have been travelling for over forty years, having explored all the giant planets in the outer solar system, including many of their moons. However, these spacecraft are also noteworthy for something else. Both carry an extraordinary disc: a disc that aims to summarise our species, encapsulate who we are – a message to any civilisations that may one day discover *Voyager 1* or *2* as they fly on their journey through outer space. A journey that would also involve John Lennon, a Grammy and a little help from Bach.

Launched in August and September of 1977, the Voyager mission was primarily focused on studying Jupiter and Saturn, including Saturn's rings and the largest of both planets' moons. Despite the initial assessment that an onward mission beyond Saturn and Jupiter would be too expensive, the mission was extended to cover Neptune, Uranus and now continues its journey into interstellar space.[1] The infrequent alignment of the gas giants allowed the Voyager mission to use the gravity of each planet to slingshot onto the next planet – a hitchhiker catching a ride onto its next destination.

The team behind the Voyager mission always knew that the spacecraft were going to leave the solar system at some point in the future, so this mission offered a unique and rare opportunity to send the craft with a message. A message to say, 'This is who we are.' Imagine you had to decide how to communicate a true reflection of our species in a simple message – what would you include? Would you describe our tendency to wage war as well as our capacity for love

and community? Which great works of art would you select and from what creative discipline? Exactly what information best (and most honestly) describes who we are?

The task was overseen by the much-loved astronomer Carl Sagan, alongside a team of brilliant minds who brought the message to life in just a few months. It was decided, following the suggestion of the project's technical director and astronomer-astrophysicist Frank Drake, that the message should take the form of a golden phonograph record. Each record was enveloped in an aluminium case along with a record needle and cartridge.

The record included a mixture of images, Earth sounds, music and greetings to give as broad a representation of our planet and its inhabitants as possible. An intergalactic dating profile, if you will, professing our best qualities. The images include competing athletes, DNA, our solar system, diagrams of conception and foetal development, a nursing mother, a classroom, a supermarket, architecture, transport, and many others.[2] The sounds include earthquakes, thunder, a chimpanzee, a train, wind, rain and laughter.[3] The music, which could probably inspire more debate on Planet Earth than any other creative discipline, included the Navajo Night Chant, 'Johnny B. Goode' by Chuck Berry, 'Dark Was the Night, Cold Was the Ground' by Blind Willie Johnson, panpipes from the Solomon Islands and three pieces of music from Bach.[4]

Tim Ferris, producer of the Golden Record, told the *New Yorker* that he sought John Lennon's help with the project, but the former Beatle had to leave the country for tax reasons and was therefore unable to oblige. But he did make a very physical impression on the disc. Ferris was inspired by Lennon's habit of etching messages into a record's 'runout groove' and decided to add a message of his own to the Golden Record which said, 'To the makers of music — all worlds, all times.'[5] The greetings on the disc were recorded in fifty-five different languages and included a variety of messages, as participants were given freedom to say what they wanted, so long as it was brief and that it was a greeting intended for extraterrestrials. One message, in Swedish, said, 'Greetings from a computer programmer in the little town of Ithaca on the Planet Earth,' while another in Amoynese said, 'Friends of space, how are you all? Have you eaten yet? Come visit us if you have time.'[6]

For many years, members of the public weren't able to enjoy the golden disc's contents in their entirety, but in 2017 the Ozma record label set up a Kickstarter to create a 'home use' version of the Golden Record containing all the content that appeared on the original to celebrate its fortieth anniversary. The Kickstarter received nearly 11,000 supporters and raised over $1.3 million. In an interview with the Vinyl Factory, David Pescovitz, one of the founders of the Kickstarter campaign, described the original Golden Record as 'an inspired scientific effort and, to me, a compelling piece of conceptual art'.[7] The record went on to be released to the public and even won a Grammy, indicating just how easily conversations about the cosmos and our place in it can inspire popular culture.

After its mission to Jupiter, Saturn, Uranus and Neptune, *Voyager 1* is now over 22 billion km from Earth, making it the farthest human-made object from our planet, and its journey continues through interstellar space. Unfortunately, it'll take *Voyager 1* at least 30,000 years just to get out of the Oort Cloud,[8] so the chances of the golden disc ever being discovered is not only astronomically small but even more remote during our species' lifespan. However, though it is a sincere attempt to reach extraterrestrial intelligence, the golden disc is just as much a symbol of our species' desire for exploration, contact and perhaps, most importantly, meaning. One of the messages on the disc was from US President Jimmy Carter, which said, 'This is a present from a small distant world, a token of our sounds, our science, our images, our music, our thoughts, and our feelings. We are attempting to survive our time so we may live into yours.'[9]

'There is no problem in science that can be solved by a man that cannot be solved by a woman.'

VERA RUBIN

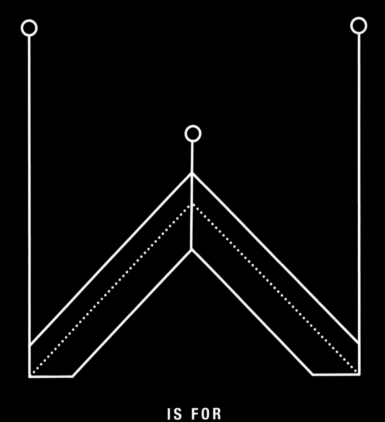

IS FOR

Women and Space

Due to a myriad of factors involving systemic sexism, archaic perceptions of gender roles and lack of opportunities, the history of discovery and progress in all matters space is often associated with male success. In 2021, according to the International Organisation of Standardisation, only 22 per cent of people employed in the international space industry identify as female,[1] highlighting that when it comes to equal representation, there's still much work to be done. This chapter is dedicated to talking about some of the pioneering women who not only contributed to scientific progress but did so in a society that often discouraged and suppressed women's participation in science. Since this book is intended to be a succinct introduction to each topic, I've only been able to talk about a small collection of these scientists that I came across during my research for the book; a richer and more comprehensive account of women's space-related discovery would require a much larger list of names and far greater level of detail to do it any form of justice. However, I hope this chapter goes some way to remind us that denying equal opportunity on the grounds of gender, sexuality or race is not only morally abhorrent, but will also serve to rob society of the potential for countless groundbreaking discoveries in the future.

Jocelyn Bell Burnell is an astrophysicist who, in 1967, discovered the first pulsars after analysing the repeating signals given off by their rapid rotations.[2] Pulsars are a type of neutron star, born when a massive star collapses in on itself, rapidly spinning and sending pulsating beams of radiation across space like a lighthouse.[3] This discovery resulted in a Nobel Prize, but surprisingly, not for Jocelyn. The prize went to her supervisor, Antony Hewish. However, Bell Burnell does feel that the prize was appropriately given to Hewish, given that she was a Ph.D. student at the time.[4] Bell Burnell commented more broadly on her experience of being a woman in science during an interview with *National Geographic*: 'The picture people had at the time of the way that science was done was that there was a senior man – and it was always a man – who had under him a whole load of minions, junior staff, who weren't expected to think, who were only expected to do as he said.'[5] Despite missing out on the Nobel Prize, Bell Burnell was awarded the Special Breakthrough Prize in Fundamental Physics in 2018 for her pioneering work. Speaking to Louise Walsh from the University of Cambridge, she explained how the £2.3 million prize money would help 'encourage a more diverse research student population from under-represented groups'.[6]

Nancy Grace Roman is commonly known as 'the mother of Hubble' for her role in bringing the world's most famous and groundbreaking telescope to life. Thankfully the discouragement she received as a young woman did nothing to dissuade her. In a series of interviews that she gave to NASA, she recalled her college's head of physics saying to her, 'You know, I usually try to talk women out of going into physics. But I think maybe – you might make it.' Even in high school, Roman's guidance counsellor responded to her request to continue studying algebra rather than Latin with, 'What lady would take mathematics instead of Latin?'[7] Nancy Roman, who was also NASA's first female executive and first chief astronomer, sadly died in 2018, but her legacy lives on in many ways. In 2020 NASA announced the launch of its most advanced and powerful telescope so far, a telescope that will develop our understanding of dark energy and the Universe. The telescope, set to launch in 2025, has been named the Nancy Grace Roman Space Telescope in honour of all her contributions to the field of astronomy. In 2019 President Donald Trump attempted to kill off the telescope as part of a wider cost-cutting proposal, but Congress later reinstated it.[8] NASA's

Dr Thomas Zurbuchen said, 'Dr Roman deserves to be permanently associated with this amazing mission that she really helped enable in a direct fashion ... I'm so delighted to have that name as a lasting legacy to this amazing person – she deserves a place in the heavens.'[9]

If we travel back to the Harvard College Observatory in 1912, there was an astronomer by the name of Henrietta Swan Leavitt who made a remarkable discovery that would change the way we view the Universe. Leavitt was one of a number of women astronomers known as the 'Harvard computers', described as such for their ability to do great analytical work, but this was a time when women were prohibited from using the telescopes and were paid a meagre wage, similar to that of unskilled workers, despite some being astronomy graduates.[10] Women still didn't have the right to vote in the US and would be denied it for another eight years. Despite these barriers, Leavitt made a remarkable discovery regarding the correlation between the frequency of a certain type of star's pulsating light and how far away they were from Earth. This made these types of stars the perfect cosmic co-ordinate for measuring the enormous distances in space. This discovery had a profound impact on our understanding of the stars, paving the way for Hubble's work on the size and expansion of the Universe. Sadly, Leavitt's huge achievements weren't celebrated fully during her lifetime, but her legacy lives on with her findings now commonly referred to as Leavitt's Law.[11]

Imagine discovering something that cannot be seen. That's exactly what astronomer Vera Rubin did in the 1970s and 1980s when her observations provided powerful evidence for the existence of dark matter – the invisible component that permeates galaxies and stops them falling apart. Rubin and her team set out to assess the distribution of mass in galaxies with the expectation that most of the mass would be centralised. This hypothesis would result in the centre of a galaxy moving more quickly than the outskirts. However, that wasn't the case. Rubin saw that the outer regions were moving way more quickly than expected, suggesting that there must be an invisible force as the cause.[12] It's now widely accepted that 26.8 per cent of the Universe's matter is, in fact, dark matter – although it's still unknown exactly what dark matter is.

Despite pursuing astronomy at a time when some universities didn't accept women (she was rejected from Princeton, which didn't accept women until 1975),[13] Rubin's body of work provided a seismic change in our understanding of the composition of the Universe. Though she passed away in 2016, Rubin remains an inspiration, never allowing barriers to prevent her immense talents from burning brightly. She once said, 'There is no problem in science that can be solved by a man that cannot be solved by a woman.' Rest in peace, Vera Rubin.[14]

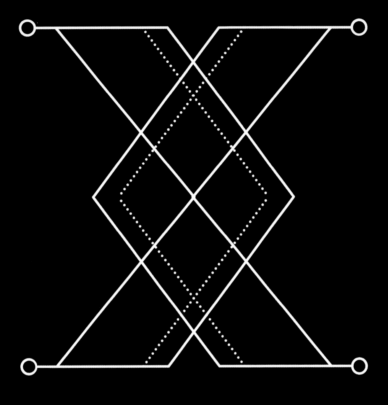

IS FOR

X-Risk

Anyone who has seen the videos of Boston Dynamics robots dancing to the Contours' 'Do You Love Me' – a different bot appearing to escape from a building, and a platoon of swan-necked arachnoid dog-bots doing press-ups[1] – has most likely felt a familiar cold sweat of existential terror.

It's normal to contemplate one's demise, be it individually or as a species, but we do seem to have a morbid fascination with exploring it in detail. Take the medium of cinema – *Armageddon, Deep Impact, Greenland, Children of Men, A Quiet Place, Geostorm, Independence Day, The Day After Tomorrow, 28 Days Later, Mad Max, Melancholia, Knowing, Sunshine, Seeking a Friend for the End of the World, The World's End, This Is the End* – the ways we have cinematically explored the imminent end to our existence is prolific and varied, from meteors to aliens, climate change to viruses. Keep calm and carry on.

X-risks, also known as existential risks, are potential threats to our very existence. They are variables that may emerge here on Earth or originate from outer space that could, at some point in the future, render humanity, and maybe even all life on our planet, extinct. If you're already prone to anxiety and hysterical reactions to worrying information, I'll calmly summarise for you, so you don't have to read the whole chapter: *WE'RE ABSURDLY FRAGILE BEINGS VULNERABLY PERCHED ON A HIGH-SPEED SPACE ROCK... THERE ARE POTENTIAL THREATS EVERYWHERE AND WE'RE PROBABLY ALL GOING TO DIE SOON. SO GO AND BUY LOTS OF TINNED FOOD AND TOILET PAPER NOW AND MOVE PERMANENTLY INTO YOUR BASEMENT!*

Although there are no immediate dangers approaching from outer space, let's start with some of the risks that could pose a danger to us from the cosmos. A fairly effective way to wipe out an entire species would be from a high-speed asteroid colliding in a violent and ungracious fashion with the Earth's surface. Evidence suggests that this is what happened 65 million years ago – which resulted in the end of the dinosaurs' 165-million-year dominion on Earth.[2] Luckily, most asteroids are burned up in Earth's atmosphere, but if a big enough asteroid collided with our planet, it could theoretically wipe out a great deal of life on Earth. The immediate impact itself would kill hundreds of thousands of people and huge amounts of debris would be thrown up into the atmosphere, blocking light. Earthquakes would be triggered across the planet, with giant tsunamis battering coastlines. And that's before talking about the acid rain.[3] But fear not, NASA has a department dedicated to monitoring 'near Earth objects' to assess their potential threat to the planet, and they're even developing technologies to deflect dangerous approaching asteroids away from Earth.[4] So you can hold off on moving your things into the basement. For now. I mean, it wouldn't really help much against that kind of threat anyway.

Aside from hypothetical space rocks hurtling through the solar system towards us, there is also the possibility that hostile alien life forms may exist elsewhere in the Universe – life forms who practise their own particular style of galactic colonisation, either for fun, subjugation or resources. Or something else. Who knows what alien imperialists are into. However, given there's no evidence of life in space so far, and certainly no evidence of aliens with a vendetta against us, this risk of imminent danger could easily be discounted as slim to none.

Unsurprisingly, in terms of existential threats, humanity's most potent enemy is a little closer to home. The scenarios that could one day present a very real danger to human survival are way more likely to come as a result of our choices and behaviour as a species. The Centre for the Study of Existential Risk (CSER) at Cambridge University focuses primarily on those existential risks that are a result of emerging technology. It sees the twenty-first century as a pivotal time in human history, where the exponential increase in technological developments must be balanced with managing the potential threats that accompany these advances. Martin Rees, one of CSER's founders, likened the studying of such risks to taking out insurance on your home – the risks might be unlikely, but their impact is potentially so serious, it's prudent to plan for how to deal with them. In a 2014 TED Talk, Rees refers to air travel having the capacity to spread a virus globally in a matter of days – something which has sadly been proven to a tragic and devastating level with Covid-19. He commented, 'Over nearly all of Earth's history, threats have come from nature – disease, earthquakes, asteroids and so forth. But from now on, the worst dangers come from us.'[5]

One of the most headline-grabbing 'risks' is related to mistrust of artificial intelligence. A number of high-profile figures, including Elon Musk and the late Stephen Hawking publicly aired their growing concerns about AI. Hawking was quoted on the BBC as saying, 'The development of full artificial intelligence could spell the end of the human race,'[6] and at the South by Southwest conference, Musk warned that it is our 'biggest existential threat' and more dangerous than nuclear weapons.[7] Many working within the AI industry would argue that the catastrophic reactions to AI are overstated, with Demis Hassabis, founder of AI research laboratory DeepMind, saying to the *Guardian*, 'We're still decades away from anything like human-level intelligence.'[8] Although AI has the potential to be

transformative to society, whether that be through data analysis or research into diseases and medicine,[9] it does seem reasonable that any technology that has the capacity to influence society in such a profound fashion needs to be observed and researched to assess the potential risks with the appropriate regulations put in place.

Whether it's doing irrevocable damage to our environment, nuclear war or becoming an organic battery in a *Matrix*-style digital dictatorship, it seems ironic that the main threats to humanity's future are of our own creation and potentially still within our control. The chances of you and I being alive at this exact moment in time are infinitesimally small. The billions of years it has taken for our species to evolve and flourish to a point where we can ponder our own existence is nothing short of a miracle: a gift to be held tightly as we fly through space.

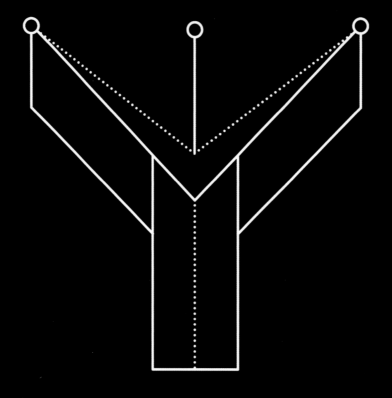

IS FOR

Yellow Dwarf

While a yellow dwarf might sound like something from Tolkien mythology, in fact, it's the reason we're all alive today. It's the reason our planet has evolved with the right conditions to cultivate life, the reason we have night and day and, most importantly, the reason we have beautiful holiday photos for Instagram. To put it more simply, our very own Sun is a yellow dwarf.

Our Sun has been on the receiving end of several cruel blows to its status over the centuries. Heralded and worshipped as a god by the ancient Egyptians, placed at the centre of the Universe by heliocentrism, it was then rather mercilessly identified as just one of many billions of stars in our galaxy alone, and a rather average-sized one at that. All things considered, this averagely sized dwarf is still the reason we're all alive today, so let's go easy on the sizeist language, OK?

The term 'yellow dwarf' refers to a star in its main sequence, whereby it converts hydrogen to helium through nuclear fusion in its core. A yellow dwarf star in its main sequence has already progressed through the early stages of stellar formulation – the necessary majestic matrimony of space dust and gas, collapsing into a hot ball, setting the stage for the stellar nuclear-fusion light show that will burn brightly for billions of years. Despite being called a yellow dwarf, it is yellow only in name, as its appearance to us is misleading – it's actually white in colour, and merely appears yellow through Earth's atmosphere.[1]

Our Sun, a typical yellow dwarf, has a circumference of just over 4.3 million km, and if you were to try and fill the Sun up with Earths, you'd need over a million of our planet to do it.[2] Now that certainly sounds big, but as we've discussed, our Sun is of average size – there are other stars out there that are much, much bigger. Near the centre of the Milky Way resides a red supergiant called UY Scuti, which has a radius 1,700 times that of our Sun.[3] It's safe to say that in space, there's always something bigger.

While we depend upon the Sun for our balmy summers and doses of vitamin D, we're lucky to be 240 million km away, as the surface temperature is around 5,500° C (10,000° F) rising to 15 million degrees Celsius (27 million degrees Fahrenheit) in its core.[4] Despite the omnipresence of our Sun, there is still much we don't know about its inner machinations. However, in January of 2020, a solar telescope positioned on a Hawaiian volcano produced the most detailed images of our Sun's surface ever taken. The first images from the Daniel K. Inouye Solar Telescope revealed bubbling plasma cells, each one about the size of Texas.[5] This bubbling is the result of convection, seeing the hot plasma cells from within the Sun rising to the surface, cooling and then returning to the interior. Using the world's largest solar telescope mirror, the Inouye Solar Telescope observes the magnetic fields of the Sun, which are the source of coronal mass ejections and solar flares. This is important, as coronal mass ejections and solar flares can impact Earth in a multitude of negative ways, including disruption to satellites, communication systems and widespread electrical malfunction.[6] Gaining a deeper understanding of how the Sun works will help us to predict these solar behaviours to mitigate against their effects on Earth. The European Space Agency's Solar Orbiter laboratory will also contribute to uncovering more of the Sun's secrets. The orbiter's complex instruments and ability to take closer imagery of the Sun than ever captured before will hopefully help us answer questions about what heats the upper layer of the Sun's atmosphere to millions of degrees, what powers the solar winds and how the Sun's magnetic activity cycle works, which sees the north and south poles flip entirely every eleven years.[7]

But what of the future of our Sun? Once a yellow dwarf has exhausted all its hydrogen, it will move on to the dying phase of stellar evolution and become a red giant. This happens when the outward pressure of nuclear fusion gives way to gravity, making the star smaller, denser and hotter; the resulting energy from helium fusion causes the star to then expand outwards to many, many times the size of the original star. This will happen to our Sun in about 5.4 billion years, so no need to pack your bags just yet. While red giants can be up to 1,000 times larger than our Sun (and red supergiants are even bigger), the surface temperatures are much cooler, as they are spread over a larger surface area, giving them a red hue – hence the name red giant.[8]

As much as our Sun is the very reason we exist today, it will also be the cause of our planet's demise and, unless we find a way to leave Earth, the destruction of our species (if we haven't destroyed ourselves first). Once it has transformed into a red giant, its outer layers will have expanded, engulfing Mercury, Venus and very likely, Planet Earth. Like Goya's *Saturn Devouring His Son*, this once verdant planet will be consumed in the Sun's red glow, and whatever remains on Earth will be returned to the cosmos as space dust in a solemn finale to our planet's life cycle.

After obliterating any semblance of life, this red giant will begin shedding its outer layers, with the ejected gas glowing elegantly as it escapes, creating what is known as a planetary nebula. Once the outer layers have drifted away into the cosmos, the remaining core of carbon and oxygen is typically about the size of Earth and known as a white dwarf.[9]

It's speculated that after many billions of years, white dwarfs will eventually transform into black dwarfs, where they would have cooled to such an extent that they emit no heat or light. However, the Universe is still way too young for any of these to exist, as they take trillions of years to form. According to theoretical physicist Matt Caplan, some of these black dwarfs with enough mass will eventually explode into supernovae due to nuclear fusion from quantum tunnelling. The fusion would create iron, which would trigger the explosions; a series of fiery swansongs long after the rest of the Universe has turned black. 'It will be a bit of a sad, lonely, cold place ... where the Universe will be mostly black holes and burned-out stars.' However, Caplan says that our own Sun does not have the required mass to result in such a supernova. 'Even with very slow nuclear reactions, our Sun still doesn't have enough mass to ever explode in a supernova, even in the far far future. You could turn the whole Sun to iron and it still wouldn't pop.'[10]

It seems a lamentable fact that something so inexorably powerful, which has burned so furiously and shone down over this planet for billions of years, will one day extinguish its final embers for ever. A lonely, ancient megalith, an iron giant, who once created, then devoured, then roams through space in darkness.

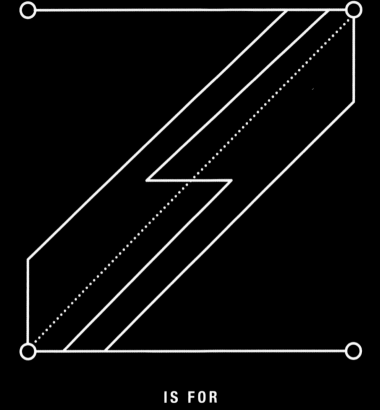

IS FOR

Zoo Hypothesis

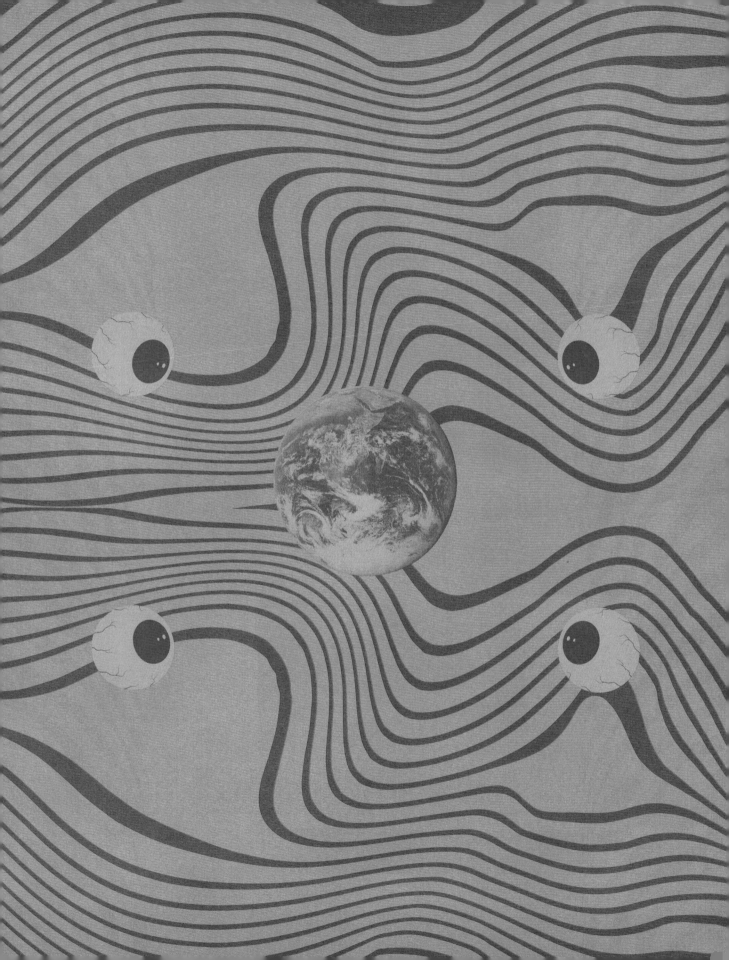

Earlier in the book, we talked about the Fermi paradox, and the question 'Where is everyone?' If the Universe is so big, so old, and there are billions of planets that should be able to support life, then where are all the cosmic companions looking for intergalactic chit-chat? The 'zoo hypothesis' takes a slightly different approach to the subject: the question isn't 'Where is everyone?' rather, 'Why haven't they said hello?'

The zoo hypothesis tackles the Fermi paradox head-on, proposing that there are not only many aliens out there, but they are fully aware of us – they've just made the decision not to communicate. There are numerous reasons why we might be receiving the cosmic cold shoulder from our intergalactic neighbours, ranging from not being of much interest to them, not being ready for them, or perhaps because the aliens are taking a hands-off approach for other reasons entirely.

Although the concept of alien civilisations deciding not to contact humans has been discussed many times throughout history, the term 'zoo hypothesis' was first mentioned by astronomer John Ball in his 1973 book of the same name.[1] In this hypothesis, alien life may well be ubiquitous across the Universe, but the reason we have yet to be contacted is that we are being observed by beings who have decided that observation rather than communication is currently the preferable course of action. Aliens who are observing us might feel that we have not as yet reached a stage in our development as a species where we are equipped to deal with the ramifications of being one of many civilisations in the Universe, a bit like a parent withholding information about Father Christmas from their small children for fear of emotional meltdowns and ruined festivities.

Despite our ability as a species to contemplate our own existence, to look back in time to the origins of the Universe and measure ripples in space-time through our scientific ingenuity, it's easy to see on many levels just how primitive we still are. There is a chance that aliens may watch us with the same interest we have in microbes – we are interesting, but they're in no rush to invite us out for coffee. In a video for *Business Insider*, Neil deGrasse Tyson said, 'I wonder if, in fact, we have been observed and upon close examination of human conduct and human behaviour they have concluded there is no sign of intelligence.'[2] After all, despite our numerous technological achievements as a species, imagine being an alien and visiting Earth between 2016 and 2021 and seeing President Trump stepping off a private jet with his phone in one hand and the nuclear codes in another. I, too, would deem us unworthy of conversation.

Other scientists have mused on the existence of a galactic club:[3] an interstellar soirée, a cosmic conglomerate of advanced civilisations who have unilaterally decided to allow earthlings to evolve independently of interference... under their watchful eye. *Star Trek*'s Prime Directive (which is often referenced in the context of this club, for obvious reasons) states, 'The Prime Directive prohibits Starfleet personnel and spacecraft from interfering in the normal development of any society, and mandates that any Starfleet vessel or crew member is expendable to prevent violation of this rule.'[4] Perhaps a galactic club would only make themselves known if we were on the brink of self-destruction, which at times seems to be a distinct possibility, given our masochistic tendencies. Alternatively, we might well be an alien civilisation's light entertainment, a planetary equivalent of *The Truman Show*, where global news is another species' end-of-season finale.

Scientist and futurist João Pedro de Magalhães thinks that it's vital for us to be actively trying to contact intelligent civilisations, effectively informing them that we are ready for contact. He thinks that by taking a proactive approach to engagement with a little interstellar lobbying, we may well be able to influence their decision-making process and perhaps engage with us sooner. He does, however, concede that there might be biological considerations at hand that may render meaningful contact difficult, such as our lifespans being a fraction of theirs.[5] This would be like us trying to negotiate a mutually beneficial Brexit agreement with a mayfly.

Others think that actively advertising our existence in the cosmos could be akin to placing your head in the open mouth of a famished crocodile. It is impossible for us to anticipate the objectives of an extraterrestrial intelligent species, so extending our hand in peace may be rewarded with a less than desirable response. In the documentary *Stephen Hawking's Favorite Places*, Hawking said, 'Meeting an advanced civilisation could be like Native Americans encountering Columbus. That didn't turn out so well.'[6]

For those of you who were about to fill your local MP's inbox with 'aliens aren't welcome in my parish' emails, it's too late. We've already begun sending messages with the specific intention of letting ETs know we're here. The practice is known as METI, short for Messaging to Extraterrestrial Intelligence, and 2017 saw Norwegian scientists beaming a radio transmission comprising music and maths towards a star twelve light years away in the hope of reaching alien ears.[7] When interviewed by the *Financial Times'* John Thornhill, the president of METI International, Douglas Vakoch, said, 'Fear does not protect us, it only restricts us,' with any aliens capable of covering interstellar space unlikely to be in need of anything material from us.[8] Mike Matessa, a cognitive scientist who works with Vakoch on what messages METI should transmit, told *WIRED* magazine, 'How can

you talk about altruism or kindness or other human qualities, and how do you build that up from math … It's advanced storytelling using languages that any civilization can understand.'[9]

Even without METI-based activities, our TV broadcasts have been leaving the planet for decades. Perhaps there is a higher intelligence out there who has no interest in engaging with us conversationally but is currently on season four of *Selling Sunset.*

The global cognitive shift that would accompany us realising that we aren't the only life in the cosmos would be far more profound than anything experienced before. While we might optimistically hope that alien civilisations would come in peace, even discovering they exist would have irrevocable ramifications on society, religion and how we interact with each other as a species – many of the national divides and disagreements we experience today would disappear when confronted with a revelation of such magnitude. For now, limited to the parameters of our existing technology, it's most likely we're in for a long wait.

'Remember to look up at the stars and not down at your feet. Try to make sense of what you see and wonder about what makes the Universe exist. Be curious. And however difficult life may seem, there is always something you can do and succeed at. It matters that you don't just give up.'

STEPHEN HAWKING

—QUOTES

'The Universe is full of magical things ...': Eden Phillpotts, A Shadow Passes, via Susan Ratcliffe (ed.), *Oxford Essential Quotations*, Oxford University Press, 2016.

'The black hole teaches us that space can be crumpled ...': John Wheeler, via Denis Overbye, 'John A. Wheeler, Physicist Who Coined the Term "Black Hole", Is Dead,' *New York Times*, 14 April 2008, www.nytimes.com/2008/04/14/science/14wheeler.html

'If you feel you are in a black hole ...': Stephen Hawking, Nordic Institute for Theoretical Physics Hawking Radiation Conference, Press Association, *Guardian*, 27 August 2015, www.theguardian.com/science/2015/aug/25/black-holes-way-out-stephen-hawking#:~:text=t%20give%20up.-,There's%20a%20 way,'&text=All%20is%20not%20lost%20if,universe%2C%20 according%20to%20Stephen%20Hawking.

'The Universe is a pretty big place ...' : Carl Sagan, *Contact*, Simon & Schuster, 1985.

'Going to the Kuiper Belt is like an archaeological dig ...': Alan Stern, via Leonard David, 'On Pluto Time: Q&A with New Horizons Leader Alan Stern, Space.com, 27 October 2015, www.space.com/30934-pluto-new-horizons-alan-stern-interview.html

'From out there on the Moon, international politics look so petty...': via Ellen Stewart, 'One incredible quote from Edgar Mitchell, the 6th man to walk on the moon', Indy100, 8 February 2016, www.indy100.com/celebrities/one-incredible-quote-from-edgar-mitchell-the-6th-man-to-walk-on-the-moon-7290476

'Although quantum mechanics is an extraordinarily successful scientific theory ...' : Jim Baggott, 'Everything you need to know about quantum physics (almost)', *Science Focus*, 5 August 2020, www.sciencefocus.com/science/quantum-physics

'The cosmos is within us ...': Carl Sagan, 'The Shores of the Cosmic Ocean', *Cosmos: A Personal Voyage*, TV, PBS, 28 September 1980.

'There is no problem in science ...': Vera Rubin, Mothers in Science, www.mothersinscience.com/trailblazers/vera-rubin

'Remember to look up at the stars ...': Stephen Hawking, via Michael McGowan, '"Remember to look up at the stars": the best Stephen Hawking quotes', *Guardian*, 14 March 2018, www.theguardian.com/science/2018/mar/14/best-stephen-hawking-quotes-quotations

—A

1. 'Cave paintings reveal use of complex astronomy', University of Edinburgh, 27 November 2018, www.ed.ac.uk/news/2018/cave-paintings-reveal-use-of-complex-astronomy

2. Joanna Gillan, 'Nabta Playa and the Ancient Astronomers of the Nubian Desert', Ancient-Origins.net, 3 August 2018, www.ancient-origins.net/ancient-places-africa/nabta-playa-and-ancient-astronomers-nubian-desert-002954

3. Kenneth Chang, 'Signs of Modern Astronomy Seen in Ancient Babylon,' *New York Times*, 28 January 2016, www.nytimes.com/2016/01/29/science/babylonians-clay-tablets-geometry-astronomy-jupiter.html; Joel Achenbach, 'Clay tablets reveal Babylonians discovered astronomical geometry 1,400 years before Europeans', *Washington Post*, 28 January 2016, www.washingtonpost.com/news/speaking-of-science/wp/2016/01/28/clay-tablets-reveal-babylonians-invented-astronomical-geometry-1400-years-before-europeans/; 'Babylonian Astronomy', Wikipedia, Last updated: 3 September 2022, en.wikipedia.org/wiki/Babylonian_astronomy

4. Christopher Minster, 'Ancient Mayan Astronomy', Thought.com, 24 July 2019, www.thoughtco.com/ancient-maya-astronomy-2136314

5. Jamie Carter, 'What is astronomy? History and Definition', Space.com, 29 October 2021, www.space.com/16014-astronomy.html; 'Cosmos,' Astronomy, Swinburn University, astronomy.swin.edu.au/cosmos/a/Astronomy; Observational and Theoretical Astronomy, Nature 131, 648 (1933), doi.org/10.1038/131648a0

6. 'Astronomy', NASA, 9 July 2020, www.nasa.gov/audience/forstudents/k-4/dictionary/Astronomy.html

7. 'NASA Astrophysics', Science.nasa.gov, science.nasa.gov/astrophysics

8. Frank H. Shu, 'Cosmology', *Encyclopaedia Britannica*, 3 December 2021, www.britannica.com/science/cosmology-astronomy

9. Gareth Willmer, 'Extremophiles could hold clues for climate change-tackling technologies', European Commission, 10 June 2021, ec.europa.eu/research-and-innovation/en/horizon-magazine/extremophiles-could-hold-clues-climate-change-tackling-technologies; 'Extremophile', Biology Dictionary, 26 May 2017, biologydictionary.net/extremophile/; Lyndsy Gazda and Smithsonian Ocean Team 'The Microbes That Keep Hydrothermal Vents Pumping', *Smithsonian*, March 2016, ocean.si.edu/ecosystems/deep-sea/microbes-keep-hydrothermal-vents-pumping

10. 'Extragalactic Astronomy', Harvard University Center For Astrophysics, www.cfa.harvard.edu/research/science-field/extragalactic-astronomy

—B

1. 'The Country Parson Who Conceived of Black Holes', *Cosmic Horizons Curriculum Collection*, ed. Steven Soter and Neil deGrasse Tyson, www.amnh.org/learn-teach/curriculum-collections/cosmic-horizons-book/john-michell-black-holes

2. 'Understanding gravity – warps and ripples in space-time', Australian Academy of Science, www.science.org.au/curious/space-time/gravity

3. 'Press Release: The Nobel Prize for Physics, 2020', Nobel Prize, 6 October 2020 www.nobelprize.org/prizes/physics/2020/press-release

4. Kurzgesagt – In a Nutshell, 'Black Holes Explained – From Birth to Death', 15 December 2015, www.youtube.com/watch?v=e-P5IFTqB98

5. John Wheeler and Kenneth Ford, *Geons, Black Holes & Quantum Foam: A Life In Physics*, W. W. Norton & Company, 1999, via Dennis Overbye, 'John A. Wheeler: Physicist Who Coined the Term "Black Hole" Is Dead at 96', New York Times, 14 April 2008.

6. 'What Is Hawking Radiation?', Science Alert, www.sciencealert.com/hawking-radiation

7. Nola Taylor Redd, 'The Beginning to the End of the Universe: How black holes die', *Astronomy*, 3 February 2021, astronomy.com/magazine/news/2021/02/the-beginning-to-the-end-of-the-universe-how-black-holes-die

8. 'Oxford Mathematician Roger Penrose jointly wins the Nobel Prize in Physics', Ox.ac.uk, 6 October 2020, www.ox.ac.uk/news/2020-10-06-oxford-mathematician-roger-penrose-jointly-wins-nobel-prize-physics

9. Marina Koren, 'What Earth Owes to Black Holes', *The Atlantic*, 7 October 2020, www.theatlantic.com/science/archive/2020/10/what-black-holes-bring-to-the-galaxy/616631

10. Elizabeth Landau, 'Black Hole Image Makes History; NASA Telescopes Coordinated Observations', NASA, 10 April 2019, www.nasa.gov/mission_pages/chandra/news/black-hole-image-makes-history

11. Ryan Whitwam, 'It Took Half a Ton of Hard Drives to Store the Black Hole Image Data', ExtremeTech, 11 April 2019, www.extremetech.com/extreme/289423-it-took-half-a-ton-of-hard-drives-to-store-eht-black-hole-image-data

12. Stuart Wolpert, 'Einstein's General Relativity Theory Beginning to Fray at the Edges', Sci Tech Daily, 25 July 2019, scitechdaily.com/einsteins-general-relativity-theory-beginning-to-fray-at-the-edges

—C

1. N. Copernicus and A. M. Duncan, *On the Revolutions of the Heavenly Spheres*, Newton Abbot, David & Charles, 1976.

2. Kevin Krisciunas, 'The Confluence of Some Ideas Used by Copernicus in De Revolutionibus', American Astronomical Society, AAS Meeting #233, id.117.06, January 2019, ui.adsabs.harvard.edu/abs/2019AAS...23311706K/abstract; 'Mu'ayyad al-Din al-'Urdi', World Science Journals, 9 November 2013, worldsciencejournals.wordpress.com/2013/11/09/muayyad-al-din-al-urdi

3. Joshua J. Mark, 'Aristarchus of Samos', World History Encyclopaedia, 16 February 2022, www.worldhistory.org/Aristarchus_of_Samos

4. Jennifer Billock, 'Visit the Site of the Biggest Witch Trial in History', *Smithsonian*, 14 September 2016, www.smithsonianmag.com/travel/visit-site-biggest-witch-trial-history-180959946/; Pam Rentz, 'The Largest Witch Hunt in World History: The Basque Witch Trials (1609–1614)', 7 October 2019, library.law.yale.edu/news/largest-witch-hunt-world-history-basque-witch-trials-1609-1614

5. 'Galileo and Astronomy', Royal Greenwich Museums, www.rmg.co.uk/stories/topics/what-did-galileo-discover

6. 'This Day In History – April 12 – Galileo Is Accused Of Heresy', History.com, 19 May 2020, www.history.com/this-day-in-history/galileo-is-accused-of-heresy; 'Galileo Galilei', Starchild, NASA/GSFC, starchild.gsfc.nasa.gov/docs/StarChild/whos_who_level2/galileo.html

7. Johannes Kepler, *Epitome of Copernican Astronomy & Harmonies of the World*, Prometheus Books, Amherst, NY, 1995.

8. Paul Setter, 'Going Bananas: The Real Story of Kepler, Copernicus and the Church', Space.com, 21 February 2017, www.space.com/35772-copernicus-vs-catholic-church-real-story.html

9. Prof. Ulinka Rublack, 'The astronomer and the witch – how Kepler saved his mother from the stake', University of Cambridge, 22 October 2015, www.cam.ac.uk/research/discussion/the-astronomer-and-the-witch-how-kepler-saved-his-mother-from-the-stake

10. Alison Flood, 'The astronomer who saved his mother from being burned as a witch', *Guardian*, 21 October 2015, www.theguardian.com/books/2015/oct/21/the-astronomer-and-the-witch-johannes-kepler-mother-katharina-witch-trial

11. Isaac Newton, *Philosophiæ Naturalis Principia Mathematica*, Apud G. & J. Innys, London, 1726.

—D

1. 'Dark Energy, Dark Matter', NASA, science.nasa.gov/astrophysics/focus-areas/what-is-dark-energy

2. Corey S. Powell, 'Darkness Invisible', *American Scientist*, www.americanscientist.org/article/darkness-invisible; Lex Clips, 'Lisa Randall: Is Dark Matter Transparent?', YouTube, 4 January 2020, www.youtube.com/watch?v=hsahmcygcMk

3. 'Jan Hendrick Oort', Your Dictionary, biography.yourdictionary.com/jan-hendrik-oort; William Harris and Craig Freudenrich, 'How Dark Matter Works', *How Stuff Works*, 16 February 2021, science.howstuffworks.com/dictionary/astronomy-terms/dark-matter.htm; John Noble Wilford, 'Jan H. Oort, Dutch Astronomer In Forefront of Field, Dies at 92', *New York Times*, 12 November 1992, www.nytimes.com/1992/11/12/us/jan-h-oort-dutch-astronomer-in-forefront-of-field-dies-at-92.html

4. Larry Sessions, 'The Coma Cluster contains thousands of galaxies', Earth Sky, 7 April 2021, earthsky.org/clusters-nebulae-galaxies/the-coma-berenices-galaxy-cluster/

5. 'Modified Newtonian Dynamics', Wikipedia, Last edited: 30 October 2022, en.m.wikipedia.org/wiki/Modified_Newtonian_dynamics

6. 'Vera Rubin and Dark Matter', *Cosmic Horizons Curriculum Collection*, www.amnh.org/learn-teach/curriculum-collections/cosmic-horizons-book/vera-rubin-dark-matter; 'Vera Rubin (1928–2016)', National Science Foundation, www.nsf.gov/news/special_reports/medalofscience50/rubin.jsp

7. 'What are WIMPS?', Universe Today, www.universetoday.com/41878/wimps

8. 'Dark Matter,' CERN, home.cern/science/physics/dark-matter

9. Bill Bryson, *A Short History of Nearly Everything*, Doubleday, London, 2004.

—E

1. 'Vesto Slipher', *Encyclopaedia Britannica*, 7 November 2021, www.britannica.com/biography/Vesto-Slipher

2. 'Expanding Universe', *Encyclopaedia Britannica*, 1 June 2021, www.britannica.com/science/expanding-universe

3. Dr William B. Ashworth, 'Scientist of the Day – Vesto Slipher', Linda Hall Library, 11 November 2016, www.lindahall.org/vesto-slipher

4. Jason Riley, 'Is the universe a bubble or a balloon?', BBC Earth, www.bbcearth.com/news/is-the-universe-a-bubble-or-balloon

5. Daniel Clery, 'Move over, Hubble: Discovery of expanding cosmos assigned to little-known Belgian astronomer-priest', Science.org, 29 October 2018, www.science.org/content/article/move-over-hubble-discovery-expanding-cosmos-assigned-little-known-belgian-astronomer

6. The Nobel Prize in Physics in Physics 2011 Press Release, 4 October 2010, www.nobelprize.org/prizes/physics/2011/press-release

7. Neil deGrasse Tyson, 'Cosmic Perspective from NOVA scienceNOW', originally broadcast 25 June 2008, www.pbs.org/wgbh/nova/sciencenow/cosmic/2009/08/dark-matter-mystery-1.html

8. Tim Childers, 'Einstein's Biggest Blunder May Have Finally Been Fixed', 6 September 2019, Live Science, www.livescience.com/solution-to-worst-prediction-in-physics.html

9. Neil deGrasse Tyson, 'Cosmic Perspective'; Cormac O'Raifeartaigh,'Investigating the legend of Einstein's "biggest blunder"', *Physics Today*, 30 October 2018, physicstoday.scitation.org/do/10.1063/PT.6.3.20181030a/full/; Britt Griswald, 'What is a Cosmological Constant', 21 December 2012, NASA 'Universe 101'; 'What is a Cosmological Constant?', National Aeronautics and Space Administration, map.gsfc.nasa.gov/universe/uni_accel.html

10. Ethan Siegel, 'Who Really Discovered the Expanding Universe?', *Forbes*, 18 November 2013, www.forbes.com/sites/startswithabang/2018/11/13/who-really-discovered-the-expanding-universe/; Cormac O'Raifeartaigh, 'Einstein's Greatest Blunder', *Scientific American*, 21 February 2017 blogs.scientificamerican.com/guest-blog/einsteins-greatest-blunder/; Arvin Ash, 'What is Dark Energy made of? Quintessence? Cosmological constant?', YouTube, 14 March 2020, www.youtube.com/watch?v=YQq0VdJApzU

11. Adam Mann, 'What is the cosmological constant?', Live Science, 16 February 2021, www.livescience.com/cosmological-constant.html; Rob Lea, 'A new generation takes on the cosmological constant', *Physics World*, 3 March 2021, physicsworld.com/a/a-new-generation-takes-on-the-cosmological-constant

12. Christopher Wanjek and Kelly Whitt, 'Quintessence: accelerating the Universe?', Astronomy Today, www.astronomytoday.com/cosmology/quintessence.html

—F

1. Elizabeth Howell, 'How Many Stars Are in the Milky Way?', Space.com, 30 March 2018, www.space.com/25959-how-many-stars-are-in-the-milky-way.html

2. 'Mission Overview: Kepler', NASA, www.nasa.gov/mission_pages/kepler/overview/index.html

3. Seth Shostak, 'Fermi Paradox', SETI Institute, 19 April 2018, www.seti.org/seti-institute/project/fermi-paradox

4. Matt Williams, 'What is the Drake Equation?', Universe Today, 13 June 2017, www.universetoday.com/39966/drake-equation-1

5. Paul Sutter, 'Alien Hunters, Stop Using the Drake Equation', Space.com, 27 December 2018, www.space.com/42739-stop-using-the-drake-equation.html

6. Nadia Drake, 'How many alien civilizations are out there? A new galactic survey holds a clue', *National Geographic*, 4 November 2020 www.nationalgeographic.co.uk/space/2020/11/how-many-alien-civilisations-are-out-there-a-new-galactic-survey-holds-a-clue

7. Daniel Brown, 'Six cosmic catastrophes that could wipe out life on earth', The Conversation, 19 January 2017, theconversation.com/six-cosmic-catastrophes-that-could-wipe-out-life-on-earth-71178

8. Irene Klotz, 'Aliens May Be Out There, But Too Distant for Contact', Space.com, 6 October 2014, www.space.com/27346-aliens-too-distant-contact.html

GENERAL READING

• Fermi's Paradox, fermisparadox.com

• Liv Boeree, 'Why haven't we found aliens yet?', VOX, 3 July 2018, www.vox.com/science-and-health/2018/7/3/17522810/aliens-fermi-paradox-drake-equation

• Ross Pomeroy, '12 Possible Reasons We Haven't Found Aliens', Space.com, 13 June 2017, www.space.com/37157-possible-reasons-we-havent-found-aliens.html

—G

1. Sir Isaac Newton, *Philosophiae Naturalis Principia Mathematica*, 1687; Julia Leyton, 'How does Gravity work', *How Stuff Works*, science.howstuffworks.com/environmental/earth/geophysics/question232.htm; The Nobel Prize in Physics in Physics 2011 Press Release, 4 October 2010, www.nobelprize.org/prizes/physics/2011/press-release

2. Laura Geggel, 'Isaac Newton's Book Auctioned for Record-Setting $3.7 Million', Live Science, 15 December 2016

3. Tom Henderson, 'Newton's Law of Universal Gravitation', The Physics Classroom, www.physicsclassroom.com/class/circles/Lesson-3/Newton-s-Law-of-Universal-Gravitation

4. Sandra May, 'What is microgravity?', NASA, 15 February 2012, www.nasa.gov/audience/forstudents/5-8/features/nasa-knows/what-is-microgravity-58.html

5. Albert Einstein, *Relativity - The Special and the General Theory*, Methuen & Co Ltd, 1964.

6. apbiolghs, 'Gravity Visualised', YouTube, 10 March 2012, www.youtube.com/watch?v=MTY1Kje0yLg

7. Adam Hadhazy, 'Relativity's Long String of Successful Predictions', *Discovery Magazine*, 26 February 2015, www.discovermagazine.com/the-sciences/relativitys-long-string-of-successful-predictions; Elizabeth Landau, '10 Things Einstein Got Right,' NASA, 29 May 2019, solarsystem.nasa.gov/news/954/10-things-einstein-got-right

8. Kip Thorne, *The Science of Interstellar*, W. W. Norton & Company, 2014.

9. Lee Billings, 'Parsing the Science of Interstellar with Physicist Kip Thorne', *Scientific American*, 28 November 2014, blogs.scientificamerican.com/observations/parsing-the-science-of-interstellar-with-physicist-kip-thorne

10. Anna Powers, 'The Theory Of Everything: Remembering Stephen Hawking's Greatest Discovery', *Forbes*, 14 March 2018, www.forbes.com/sites/annapowers/2018/03/14/the-theory-of-everything-remembering-stephen-hawkings-greatest-contribution/?sh=5cc84a7523ed

11. 'Sir Isaac Newton, Moving Words', BBC, https://www.bbc.co.uk/worldservice/learningenglish/movingwords/shortlist/newton.shtml

12. 'Albert Einstein Quotes', Notable Quotes, notable-quotes.com/e/einstein_albert.html

—H

1. Ethan Siegel, 'Who really discovered the expanding Universe?', *Forbes*, 13 November 2018, www.forbes.com/sites/startswithabang/2018/11/13/who-really-discovered-the-expanding-universe/?sh=206cda6130e8

2. Rob Garner, 'About – Hubble History Timeline. Full Text', NASA, 13 October 2020, www.nasa.gov/content/goddard/hubble-timeline-full-text; 'Scientific Uses of the Large Space Telescope', National Research Council, The National Academies Press, 1969, www.nap.edu/catalog/12399/scientific-uses-of-the-large-space-telescope

3. Christine Pullium, 'WFIRST Telescope named for "Mother Of Hubble" Nancy Grace Roman', Hubblesite, hubblesite.org/contents/news-releases/2020/news-2020-35, 20 May 2020

4. Jeff Hecht, 'The testing error that led to Hubble mirror fiasco', *New Scientist*, 18 August 1990, www.newscientist.com/article/mg12717301-000-the-testing-error-that-led-to-hubble-mirror-fiasco

5. 'Hubble's Mirror Flaw', NASA, www.nasa.gov/content/hubbles-mirror-flaw

6. 'Discoveries: Highlights of Hubble's Exploration of the Universe', NASA, www.nasa.gov/content/goddard/2017/highlights-of-hubble-s-exploration-of-the-universe

7. 'About: Webb Key facts', NASA, www.jwst.nasa.gov/content/about/faqs/facts.html

8. Natasha Pinol, Alise Fisher and Laura Betz, 'NASA's Webb Telescope Launches to See First Galaxies, Distant Worlds', NASA, 25 December 2021, www.nasa.gov/press-release/nasas-webb-telescope-launches-to-see-first-galaxies-distant-worlds

9. 'What has the Hubble telescope discovered', Royal Museums Greenwich, www.rmg.co.uk/stories/topics/what-has-hubble-space-telescope-discovered

10. 'Great Discoveries of the Hubble Space Telescope', Google Arts & Culture, artsandculture.google.com/exhibit/great-discoveries-of-the-hubble-space-telescope-nasa/KwKyaIW2Bor4IA?hl=en

GENERAL READING

• 'Hubble Space Telescope', NASA, www.nasa.gov/mission_pages/hubble/main/index.html

• Neta A. Bahcall, 'Hubble's Law and the expanding universe', PNAS, 17 March 205, www.pnas.org/content/112/11/3173

—I

1. David Bowie, 'Space Oddity', 1969, Philips, Mercury, RCA.

2. 'David Bowie' Facebook page, Facebook, www.facebook.com/davidbowie/photos/chris-hadfield-sings-space-oddity-in-spacehallo-spaceboycommander-chris-hadfield/10151372549242665

3. 'International Space Station legal framework', ESA, www.esa.int/Science_Exploration/Human_and_Robotic_Exploration/International_Space_Station/International_Space_Station_legal_framework

4. 'Visitors to the station by country', NASA, 15 November 2020, www.nasa.gov/feature/visitors-to-the-station-by-country/ (Statistics correct as of June 2020)

5. Sandra May, 'What Is Microgravity', NASA, 30 March 2010, www.nasa.gov/audience/forstudents/k-4/stories/nasa-knows/what-is-microgravity-k4.html

6. 'A Researcher's Guide to: International Space Station: Human Research', NASA ISS Program Science Office, NASA, www.nasa.gov/sites/default/files/atoms/files/np-2015-04-020-jsc_human_research-iss-mini-book-508-v2.pdf

7. 'Space Station Research Experiments', NASA, www.nasa.gov/mission_pages/station/research/experiments_category; Alexandra Witz, 'Astronauts have conducted nearly 3,000 experiments aboard the ISS', *Nature*, 3 November 2020 doi.org/10.1038/d41586-020-03085-8

8. 'Space Station Research Experiments', NASA; '15 Ways the International Space Station is Benefitting Earth', NASA, 30 October 2015, www.nasa.gov/mission_pages/station/research/news/15_ways_iss_benefits_earth

9. VideoFromSpace, 'In Space Everyone Can Hear You Poop', YouTube, www.youtube.com/watch?v=MgMYqxdVAlA

10. *Wired*, 'Former NASA astronaut Explains How Hygiene is Different In Space', YouTube, www.youtube.com/watch?v=TZkuQUCUYgM; European Space Agency, 'How Do You Shower In Space', Youtube, www.youtube.com/@EuropeanSpaceAgency

11. Channel 4, 'How Does Food Get Delivered To Space', YouTube, www.youtube.com/watch?v=u3zYG98ah04

12. NASA Johnson, 'Recycling Water on Space Station', YouTube, www.youtube.com/watch?v=womKV58QTHY&feature=emb_title

13. Andrew Griffin, 'First Ever Commercial Crew To Fly To International Space Station Revealed By Axiom Space', *Independent*, 26 January 2021, www.independent.co.uk/life-style/gadgets-and-tech/space/nasa-iss-space-station-axiom-b1792875.html; 'Commercial and Marketing Pricing Policy', NASA, www.nasa.gov/leo-economy/commercial-use/pricing-policy

14. 'Orbiting Above Earth', Axiom Space, 2020, www.axiomspace.com/axiom-station

15. 'NASA plans to take International Space Station out of orbit in January 2031 by crashing it into "spacecraft cemetery"', Sky News, 1 February 2022, news.sky.com/story/nasa-plans-to-take-international-space-station-out-of-orbit-in-january-2031-by-crashing-it-into-spacecraft-cemetery-12530194

—J

1. 'How Many Earths Could Fit Inside Jupiter?', The Nine Planets, nineplanets.org/questions/how-many-earths-can-fit-in-jupiter

2. Dr Alastair Gunn, 'Could Jupiter Become A Star', *BBC Science Focus*, www.sciencefocus.com/space/could-jupiter-become-a-star

3. Nola Taylor Tillman, 'How far away is Jupiter', Space.com, 1 June 2017, www.space.com/18383-how-far-away-is-jupiter.html

4. Phil Plait, 'BAFact Math: Jupiter is big enough to swallow all the rest of the planets whole', *Discover Magazine*, 22 August 2012, www.discovermagazine.com/the-sciences/bafact-math-jupiter-is-big-enough-to-swallow-all-the-rest-of-the-planets-whole

5. SpacePlace, 'What is a barycentre?', 3 June 2020, NASA, spaceplace.nasa.gov/barycenter/en

6. Matt Williams, 'Does Jupiter Have A Solid Core?', Universe Today, 7 May 2017, www.universetoday.com/14470/does-jupiter-have-a-solid-core

7. Michael Greshko, 'Babylonians Tracked Jupiter With Advanced Tools: Trapezoids', 28 January 2016, www.nationalgeographic.com/news/2016/01/160128-math-geometry-babylon-jupiter-astronomy-space/; Megan Gannon, 'Babylonians Tracked Jupiter with Fancy Math, Tablet Reveals', *Scientific American*, 1 February 2016, www.scientificamerican.com/article/babylonians-tracked-jupiter-with-fancy-math-tablet-reveals/; Haroon Siddique, 'Babylonians used geometry to track Jupier 1400 years before Europeans', *Guardian*, 28 January 2016, www.theguardian.com/science/2016/jan/28/babylonians-used-geometry-to-track-jupiter-1400-years-before-europeans

8. 'Jupiter's rings', Royal Museum Greenwich, www.rmg.co.uk/stories/topics/does-jupiter-have-rings

9. Kevin Grazier, 'Jupiter as a Sniper Rather Than a Shield', *Bulletin of the American Astronomical Society*, American Astronomical Society, September 2008, Vol. 40, p.404, ui.adsabs.harvard.edu/abs/2008DPS....40.1201G/abstract

10. 'Jupiter Moons', NASA, 19 December 2019, solarsystem.nasa.gov/moons/jupiter-moons/in-depth

11. Paul Scott Anderson, 'Wow! New volcano on Jupiter's moon', Earth Sky, 23 July 2018, earthsky.org/space/new-hot-spot-on-io-active-volcano

12. Tricia Talbert, 'Io Volcano Observer: Follow the Heat and Hunting Clues to Planet Evolution', NASA, 18 March 2021, www.nasa.gov/planetarymissions/io-volcano-observer; Kim Ann Zimmermann, 'Io: Facts about Jupiter's Volcanic Moon', Space.com, 14 August 2018, www.space.com/16419-io-facts-about-jupiters-volcanic-moon.html

13. Amanda Barnett, 'Europa: Ocean Moon', NASA, solarsystem.nasa.gov/moons/jupiter-moons/europa/in-depth

14. 'Europa's Interior May Be Hot Enough to Fuel Seafloor Volcanoes', NASA, 25 May 2021, europa.nasa.gov/news/32/europas-interior-may-be-hot-enough-to-fuel-seafloor-volcanoes

15. Europa Clipper, Jet Propulsion Lab, NASA, www.jpl.nasa.gov/missions/europa-clipper

16. 'Europa's Interior May Be Hot Enough to Fuel Seafloor Volcanoes', NASA, 25 May 2021, europa.nasa.gov/news/32/europas-interior-may-be-hot-enough-to-fuel-seafloor-volcanoes

—K

1. J. J. Kavelaars, 'Kuiper belt', *Encyclopaedia Britannica*, 10 November 2021, www.britannica.com/place/Kuiper-belt

2. 'Kuiper Belt', NASA, 13 November 2019, solarsystem.nasa.gov/solar-system/kuiper-belt/overview

3. James Miller, 'The Kuiper Belt in the Outer Solar System', Astronomy Trek, 21 July 2015, www.astronomytrek.com/the-kuiper-belt-in-the-outer-solar-system

4. Leo Enright, 'Pluto images boost legacy of a dogged Irish astronomer', *Irish Times*, 17 July 2015, www.irishtimes.com/news/science/pluto-images-boost-legacy-of-a-dogged-irish-astronomer-1.2287616

5. 'Kuiper Belt', NASA, 13 November 2019

6. 'Where do comets come from?', Star Child.com, December 2001, starchild.gsfc.nasa.gov/docs/StarChild/questions/question40.html

7. Armand H. Delsemme and Paul Weissman, 'Comet', *Encyclopaedia Britannica*, 6 August 2020, www.britannica.com/science/comet-astronomy

8. Suzy Stewart, 'Kuiper Belt Facts', The Planets, theplanets.org/kuiper-belt

9. Preston Dyches, '10 Things to Know About the Kuiper Belt', NASA, 14 December 2018, solarsystem.nasa.gov/news/792/10-things-to-know-about-the-kuiper-belt/

10. Dwayne Brown, George Diller and Michael Buckley, 'NASA's Pluto Mission Launched Toward New Horizons', NASA, 1 September 2006, www.nasa.gov/mission_pages/newhorizons/news/release-20060119.html

11. Tricia Talbert, 'NASA's New Horizons Reaches a Rare Space Milestone', NASA, 15 April 2021, www.nasa.gov/feature/nasa-s-new-horizons-reaches-a-rare-space-milestone; Tricia Talbert, 'NASA's New Horizons Conducts the First Interstellar Parallax Experiment', 10 June 2020, www.nasa.gov/feature/nasa-s-new-horizons-conducts-the-first-interstellar-parallax-experiment

GENERAL READING

• Rachel Kaufman, 'Three New "Plutos"? Possible Dwarf Planets Found', *National Geographic*, 12 August 2011, www.nationalgeographic.com/news/2011/8/110811-three-new-dwarf-planets-pluto-kuiper-belt-space-science

• Leonard David, 'On Pluto Time: Q&A with New Horizons Leader Alan Stern', Space.com, 27 October 2015, www.space.com/30934-pluto-new-horizons-alan-stern-interview.html

NOTES

—L

1. 'Are we all going to die next Wednesday?', *Daily Mail Online*, www.dailymail.co.uk/sciencetech/article-1052354/Are-going-die-Wednesday.html

2. Sean Martin, 'Large Hadron Collier could create BLACK HOLE and DESTROY EARTH, top astronomer claims', *Daily Express*, 1 October 2019, www.express.co.uk/news/science/1025243/black-hole-end-of-the-world-martin-rees-large-hadron-collider-particle-accelerator

3. Rachel O'Donoghue, 'Earth could be "reduced to 330ft sphere" if Large Hadron Collider experiments go wrong', *Daily Star*, 2 October 2018, www.dailystar.co.uk/news/weird-news/earth-330ft-sphere-large-hadron-16793318

4. 'Will CERN create black holes', CERN, home.cern/resources/faqs/will-cern-generate-black-hole

5. Brian Cox, 'What Went Wrong At The LHC', TED Talks, 2009, www.ted.com/talks/brian_cox_what_went_wrong_at_the_lhc/transcript?language=en

6. 'W and Z bosons', Wikipedia, Last updated: 13 September 2022, en.wikipedia.org/wiki/W_and_Z_bosons

7. 'The Birth Of The Web', CERN, home.cern/science/computing/birth-web

8. Chris Llewellyn Smith, 'Genesis Of The Large Hadron Collider', Royal Society Publishing, royalsocietypublishing.org/doi/pdf/10.1098/rsta.2014.0037

9. Alan Boyle, 'Scientists turn on biggest "Big Bang Machine"', NBC News, 10 September 2008, www.nbcnews.com/id/wbna26439957

10. 'What is the Higgs Boson and why does it matter?', ATLAS / CERN, https://atlas.cern/higgs-boson-landmark-discovery

11. Achintya Rao, 'The Higgs boson: Revealing nature's secrets', 4 July 2020, home.cern/news/series/lhc-physics-ten/higgs-boson-revealing-natures-secrets

12. 'Future Circular Collider', CERN, home.cern/science/accelerators/future-circular-collider

13. Sabine Hossenfelder, 'Particle physicists want money for bigger collider', 16 January 2019, backreaction.blogspot.com/2019/01/particle-physicists-want-money-for.html

—M

1. 'Hubble Reveals Observable Universe Contains 10 Times More Galaxies Than Previously Thought', NASA, 13 October 2016, www.nasa.gov/feature/goddard/2016/hubble-reveals-observable-universe-contains-10-times-more-galaxies-than-previously-thought

2. 'Spiral & Barred Spiral Galaxies', Space FM, www.space.fm/astronomy/starsgalaxies/spiralbarredspiral.html; Futurity.org, www.futurity.org/so-why-is-the-milky-way-a-barred-spiral

3. Joshua J Mark, 'Democritus', World History Encyclopedia, Ancient.eu., 26 May 2021, www.worldhistory.org/Democritus/; 'The Milky Way; two theories', Greek Tourist Guide, 15 June 2020, aristotleguide.wordpress.com/2020/06/15/the-milky-way-two-theories

4. 'Galileo Galilei: First To See The Milky Way Galaxy', Lumen: Introduction To Astronomy, Course Hero, https://courses.lumenlearning.com/atd-fscj-introastronomy/chapter/galileo-galilei-first-to-see-the-milky-way-galaxy

5. Sidney van den Bergh, 'Shapley and Hubble: Different Views Brought Galaxies Into Focus', *Physics Today*, 1 September 2004, physicstoday.scitation.org/doi/10.1063/1.1809079

6. Hubble Space Telescope, NASA, www.nasa.gov/mission_pages/hubble/story/the_story.html; Randy Alfred, 'Hubble Reveals We Are Not Alone', *Wired*, 30 December 1924, www.wired.com/2009/12/1230hubble-first-galaxy-outside-milky-way

7. Charles Q. Choi, 'This 3D Map of the Milky Way Is the Best View Yet of Our Galaxy's Warped, Twisted Shape', Space.com, 1 August 2019, www.space.com/milky-way-3d-map-warped-shape.html; 'The Milky Way Galaxy', NASA, 4 May 2018, solarsystem.nasa.gov/resources/285/the-milky-way-galaxy

8. Morgan McFall-Johnsen, 'NASA Scientist Shows Dinosaurs Roamed Earth On The Other Side Of The Milky Way', *Business Insider*, 8 November 2019, www.sciencealert.com/dinosaurs-roamed-the-earth-on-the-other-side-of-the-milky-way

9. 'Supermassive Black Hole Sagittarius A*', NASA, 29 August 2013, www.nasa.gov/mission_pages/chandra/multimedia/black-hole-SagittariusA.html

10. The Physics arXiv Blog, 'Black Hole Theory Finally Explains How Galaxies Form', 12 January 2022, www.discovermagazine.com/the-sciences/black-hole-theory-finally-explains-how-galaxies-form

11. 'How Many Stars Are There in the Universe', ESA, www.esa.int/Science_Exploration/Space_Science/Herschel/How_many_stars_are_there_in_the_Universe; 'One billion stars and counting - the sky according to Gaia's second data release', ESA, 25 May 2020, www.esa.int/ESA_Multimedia/Videos/2020/05/One_billion_stars_and_counting_the_sky_according_to_Gaia_s_second_data_release

12. 'Milky Way's warp caused by Galactic collision, Gaia suggests,' ESA, 2 March 2020, sci.esa.int/web/gaia/-/milky-way-s-warp-caused-by-galactic-collision-gaia-data-suggests

13. 'As many as six billion Earth-like planets in our galaxy, according to new estimates', University of British Columbia, Science Daily, 16 June 2020, science.ubc.ca/news/many-six-billion-earth-planets-our-galaxy-according-new-estimates

14. Tom Westby and Christopher J. Conselice, 'The Astrobiological Copernican Weak and Strong Limits for Intelligent Life', *The Astrophysical Journal*, 15 June 2020, iopscience.iop.org/article/10.3847/1538-4357/ab8225

—N

1. Jim Wilson, 'NASA history Overview', NASA, 2 April 2018, www.nasa.gov/content/nasa-history-overview

2. '5 Key Cold War Events', Norwich University Online, 3 July 2017, online.norwich.edu/academic-programs/resources/5-key-cold-war-events; 'The Space Race', History.com, 22 February 2010, www.history.com/topics/cold-war/space-race

3. 'Space Race', Wikipedia, Last updated: 24 October 2022, en.wikipedia.org/wiki/Space_Race; Ethan Siegel, 'This is Why The Soviet Union Lost "The Space Race" to The USA', *Forbes*, 11 July 2019, www.forbes.com/sites/startswithabang/2019/07/11/this-is-why-the-soviet-union-lost-the-space-race-to-the-usa

4. British Movietone, 'President Kennedy's Man on the Moon Speech - 1961 | Today in History', 25 May 2016, www.youtube.com/watch?v=GhgVZLrxiu0&t=91s

5. Graham Kendall, 'Would your mobile phone be powerful enough to get you to the Moon?', The Conversation, 1 July 2019, theconversation.com/would-your-mobile-phone-be-powerful-enough-to-get-you-to-the-moon-115933

6. James Jeffrey, 'Apollo 11: "The greatest single broadcast in television history"', BBC, 10 July 2019, www.bbc.co.uk/news/world-us-canada-48857752

7. Julia Layton & Mark Mancini, 'NASA's 10 Greatest Achievements', *How Stuff Works*, 9 October 2018, science.howstuffworks.com/ten-nasa-achievements.htm; Clara Moskowitz, 'NASA's 10 Greatest Science Missions', Space.com, 6 March 2009, www.space.com/6378-nasas-10-greatest-science-missions.html; Tricia Talbert, 'NASA's New Horizons Reaches a Rare Space Milestone', NASA, 15 April 2001, www.nasa.gov/feature/nasa-s-new-horizons-reaches-a-rare-space-milestone

8. '2020 Mission: Perseverance Rover', NASA, mars.nasa.gov/mars2020

GENERAL READING

• 'A Chronology of Mars Exploration', NASA, history.nasa.gov/marschro.htm

• 'NASA', Wikipedia, Last updated: 26 October 2022, en.wikipedia.org/wiki/NASA

• New Horizons, NASA, solarsystem.nasa.gov/missions/new-horizons/in-depth

• Jim Wilson, 'NASA History Overview', NASA, 2 April 2018, www.nasa.gov/content/nasa-history-overview

—O

1. 'Overview', Planetary Collective, Dir: Guy Reid, Vimeo, vimeo.com/55073825; Ivan De Luce, 'Something profound happens when astronauts see Earth from space for the first time', *Business Insider*, 16 July 2019, www.businessinsider.com/overview-effect-nasa-apollo8-perspective-awareness-space-2015-8

2. 'The Overview Effect', Houston We Have A Podcast, NASA, www.nasa.gov/johnson/HWHAP/the-overview-effect

3. London Real, 'Overview Effect Gave Me Perspective', YouTube, 5 March 2016, www.youtube.com/watch?v=p1AxBGV4WL0

4. 'A seat to fly with Jeff Bezos to space sells for $28 million,' Christian Davenport, *Washington Post*, 12 June 2021, www.washingtonpost.com/technology/2021/06/12/jeff-bezos-blue-origin-auction/; Kate Duffy, 'Jeff Bezos has said Blue Origin has sold nearly $100 million worth of tickets for future tourist trips to space', *Business Insider*, 21 July 2021, www.businessinsider.com/jeff-bezos-blue-origin-sold-100-million-space-tourism-tickets-2021-7

5. 'REWIND Send British Astronaut Tim Peake Back Into Space,' LBB Online, 13 Nov 2017; https://www.lbbonline.com/news/rewind-sends-british-astronaut-tim-peake-back-into-space

6. 'SpaceVR offers a new perspective on Earth – from a tank in Brooklyn', *Wired*, 6 July 2020, www.wired.co.uk/article/space-vr-technology

7. Edgar D. Mitchel Quotes, www.goodreads.com/quotes/416837-you-develop-an-instant-global-consciousness-a-people-orientation-an; 'Edgar Mitchell's Strange Voyage', People, 8 April 1974, people.com/archive/edgar-mitchells-strange-voyage-vol-1-no-6

—P

1. Paul Rincon, 'Why is Pluto no longer a planet?', BBC, 13 July 2015, www.bbc.co.uk/news/science-environment-33462184

2. Jamie Carter, 'Should Pluto Be A Planet Again? 13 Years After Being Demoted NASA Boss Wants "Dwarf Planet" Back', *Forbes*, 27 October 2019, www.forbes.com/sites/jamiecartereurope/2019/10/27/remember-when-pluto-was-a-planet-the-head-of-nasa-does-and-he-wants-to-wind-back-the-clock

3. Jesse Emspak, 'A Brief History of the Hunt for Planet X', *Smithsonian Magazine*, 15 December 2015, www.smithsonianmag.com/science-nature/brief-history-hunt-planet-x-180957551/?no-ist

4. 'Pluto Discovered', History.com, 3 March 2010, www.history.com/this-day-in-history/pluto-discovered

5. Preston Dyches, '10 Things To Know About The Kuiper Belt', NASA, 14 December 2018, solarsystem.nasa.gov/news/792/10-things-to-know-about-the-kuiper-belt

6. 'Why is Pluto no longer a planet?', Library of Congress, 19 November 2019, www.loc.gov/everyday-mysteries/item/why-is-pluto-no-longer-a-planet

7. 'What is a Planet', NASA, 19 December 2019, solarsystem.nasa.gov/planets/in-depth

8. Kenneth Chang, 'Pluto's Not a Planet? Only in New York', *New York Times*, 22 January 2001, www.nytimes.com/2001/01/22/nyregion/pluto-s-not-a-planet-only-in-new-york.html

9. Neil deGrasse Tyson, 'Pluto's Honour', *Natural History Magazine*, February 1999

10. Neil deGrasse Tyson, 'Astronomer Responds to Pluto-Not-a-Planet Claim', Open Letter, Space.com, 2 February 2001, www.space.com/1925-astronomer-responds-pluto-planet-claim.html

11. '6 Angry Letters Kids Sent Neil deGrasse Tyson About Pluto', Mental Floss, 4 August 2013, www.mentalfloss.com/article/52042/6-angry-letters-kids-sent-neil-degrasse-tyson-about-pluto

12. 'Pluto Moons', NASA, solarsystem.nasa.gov/moons/pluto-moons/in-depth

13. Bill Keeter, 'One Year Later: New Horizons' Top 10 Discoveries at Pluto', NASA, 14 July 2015, www.nasa.gov/feature/one-year-later-new-horizons-top-10-discoveries-at-pluto

14. Michele Debczak, 'NASA Boss Jokingly Declares Pluto A Planet, Gets Everyone's Hopes Up', Mental Floss, 30 August 2019, www.mentalfloss.com/article/598334/nasa-boss-jokingly-declares-pluto-planet

15. Loren Grush, 'Watch a debate over whether Pluto should get its planet status back', The Verge, 28 April 2019, www.theverge.com/2019/4/28/18518014/pluto-debate-planet-definition-alan-stern-international-astronomical-union

16. Amanda Kooser, 'Queen guitarist Brain May weighs in on Pluto as a planet', CNET, 29 August 2019, www.cnet.com/news/queen-guitarist-brian-may-weighs-in-on-pluto-as-a-planet/; @brianmayforreal, Instagram, 29 August 2019, www.instagram.com/p/B1u-vB7Fcml

GENERAL READING

- 'Where is New Horizons', New Horizons, pluto.jhuapl.edu/Mission/Where-is-New-Horizons.php

- Tricia Talbert, 'New Horizons: The First Mission to the Pluto System and the Kuiper Belt', NASA, 4 August 2017, rb.gy/ulu2sj

- Jim Wilson, 'NASA History Overview', NASA, 2 April 2018, rb.gy/qhkms7

—Q

1. 'Max Planck: Originator Of Quantum Theory', European Space Agency, 12 August 2012, www.esa.int/Science_Exploration/Space_Science/Planck/Max_Planck_Originator_of_quantum_theory

2. 'The Birth of Quantum Theory', History.com, 10 November 2019, www.history.com/this-day-in-history/the-birth-of-quantum-theory

3. Helge Kragh, 'Max Planck: the reluctant revolutionary', *Physics World*, 1 December 2000, physicsworld.com/a/max-planck-the-reluctant-revolutionary

4. 'Feynman: Probability and Uncertainty in Quantum Mechanics', ColdFusionNow, YouTube, 8 November 2010, www.youtube.com/watch?v=kekayfI8Ii8

5. Chad Orzel, 'What Has Quantum Mechanics Ever Done For Us?', *Forbes*, 13 August 2015, www.forbes.com/sites/chadorzel/2015/08/13/what-has-quantum-mechanics-ever-done-for-us/; 'If You Don't Understand Quantum Physics, Try This!', Domain Of Science, YouTube, 25 February 2019, www.youtube.com/watch?v=Usu9xZfabPM

6. Sean Carrol, 'Even Physicists Don't Understand Quantum Mechanics,' *New York Times*, 7 September 2019, www.nytimes.com/2019/09/07/opinion/sunday/quantum-physics.html

7. Joshua Howgego, 'Schrodinger's cat', *New Scientist*, www.newscientist.com/definition/schrodingers-cat

8. D.J.P. , 'The Economist Explains: What is spooky action at a distance?', *Economist*, 16 March 2017, www.economist.com/the-economist-explains/2017/03/16/what-is-spooky-action-at-a-distance

9. Sebastian Anthony, 'New quantum teleportation record paves the way towards a worldwide quantum network', Extreme Tech, 6 September 2012, www.extremetech.com/extreme/135561-new-quantum-teleportation-record-paves-the-way-towards-a-worldwide-quantum-network; University of Vienna, '143km: Physicists break quantum teleportation distance', Phys.org, 5 September 2012, phys.org/news/2012-09-km-physicists-quantum-teleportation-distance.html

10. minutephysics, 'How the Sun works: Fusion and Quantum Tunnelling', minutephysics, YouTube, 25 July 2011, www.youtube.com/watch?v=gS1dpowPlE8; 'Nuclear fission and fusion', BBC Bitesize, www.bbc.co.uk/bitesize/guides/zx86y4j/revision/3; Katrina Kramer, 'Explainer: What is quantum tunnelling', Chemistry World, 30 July 2020, www.chemistryworld.com/news/explainer-what-is-quantum-tunnelling/4012210.article

11. Brian Greene, 'There could be particles that travel faster than the speed of light', Tech Insider, 25 October 2006, YouTube, www.youtube.com/watch?v=GEsTmO6HvZE

12. Jim Baggott, 'What We Know About Quantum Physics', *Science Focus*, 5 August 2020, www.sciencefocus.com/science/quantum-physics

GENERAL READING

- Rafi Letzter, 'The 12 Most Important and Stunning Quantum Experiments of 2019', Live Science, 30 December 2019, rb.gy/sprqdq

- Dr Sean Carroll, 'Even Physicists Don't Understand Quantum Mechanics', *New York Times*, 7 September 2019, rb.gy/sipgqt

—R

1. Michael Greshko, 'Planet Mars, explained', *National Geographic*, www.nationalgeographic.com/science/article/mars-1

2. Tim Wallace, 'First mission to Mars: Mariner 4's special place in history', *Cosmos*, 14 July 2017, cosmosmagazine.com/space/first-mission-to-mars-mariner-4s-special-place-in-history

3. Lonnie Shekhtman, 'With First Martian Samples Packed, Perseverance Initiates Remarkable Sample Return Mission', NASA, 12 October 2021, www.nasa.gov/feature/goddard/2021/with-first-martian-samples-packed-perseverance-initiates-remarkable-sample-return-mission

4. ElderFox Documentaries, '"7 Minutes Of Terror": How Perseverance will land on Mars', YouTube, 29 July 2020, www.youtube.com/watch?v=y0_2wygoZc4; NASA, 'Perseverance Rover's Descent and Touchdown on Mars', YouTube, 22 February 2021, www.youtube.com/watch?v=4czjS9h4Fpg; 'Strongest supersonic parachute', Guinness World Records, 7 September 2018, www.guinnessworldrecords.com/world-records/648990-strongest-supersonic-parachute

5. 'Mars Helicopter Demo', NASA, mars.nasa.gov/technology/helicopter

6. 'Mars The Red Planet', NASA, 19 December 2019, solarsystem.nasa.gov/planets/mars/overview

7. 'NASA Research Suggests Mars Once Had More Water Than Earth's Arctic Ocean', NASA, 5 March 2015, www.nasa.gov/press/2015/march/nasa-research-suggests-mars-once-had-more-water-than-earth-s-arctic-ocean

8. 'NASA Finds Ancient Organic Material, Scrabbl.com, Mystery Methane on Mars', NASA, 7 June 2018, www.nasa.gov/press-release/nasa-finds-ancient-organic-material-mysterious-methane-on-mars

9. 'Organic matter found in Mars,' Joydeep Dasgupta, www.scrabbl.com/organic-matter-found-in-mars

10. Mike Wall, '2021 was an epic year for space exploration', Space.com, 31 December 2021, www.space.com/china-tianwen-1-mars-orbiter-planet-survey

11. 'Tianwen-1 and Zhurong, China's Mars orbiter and rover', Planetary Society, www.planetary.org/space-missions/tianwen-1; 'Tianwen-1', EO Portal Directory, directory.eoportal.org/web/eoportal/satellite-missions/t/tianwen-1

—S

1. 'Happy birthday to Frank Drake', SETI Institute, 25 May 2017, www.seti.org/happy-birthday-frank-drake

2. Seth Shostak, 'Project Ozma', SETI Institute, July 2021, www.seti.org/project-ozma

3. 'Has the SETI Institute found an extraterrestrial signal', SETI Institute, FAQs, www.seti.org/faq#obs1

4. Jason T Wright, 'SETI is Not About Getting Attention', State Penn University, 2018, sites.psu.edu/astrowright/2018/01/20/seti-is-not-about-getting-attention

5. Stephen J Garber, 'Searching for good science: The cancellation of NASA's SETI program', *Journal of The British Interplanetary Society*, NASA History Office, 1999, history.nasa.gov/garber.pdf

6. Seth Shostak, 'Allen Telescope Array Overview', SETI Institute, www.seti.org/ata

7. Simon Spichak, 'A scientific correction finds Venus's atmosphere probably does not contain phosphine gas', Massive Science, 30 March 2021, massivesci.com/notes/mistakes-venus-phosphine-sulfur-dioxide-astronomy

8. Seth Shostak, 'Why alien "megastructures" may hold key to making contact with extraterrestrials', SETI Institute, 24 April 2019, www.seti.org/why-alien-megastructures-may-hold-key-making-contact-extraterrestrials

9. James Wilkins & Prof. Michael Garrett, 'Are We Alone?', *DARK: Chats About Space,* podcast, 30 September 2018.

10. Jesse Emspak, 'Comet Likely Didn't Cause Bizarre "Wow!" Signal (But Aliens Might Have)', Live Science, 12 June 2017, www.livescience.com/59442-astronomers-skeptical-about-wow-signal.html

11. 'Protocols for an ETI Signal Detection', SETI Institute, 23 April 2018, www.seti.org/protocols-eti-signal-detection

12. Michael Rundle, '$100m Breakthrough Listen is 'largest ever' search for alien civilisations', 28 July 2015, *Wired*, www.wired.co.uk/article/breakthrough-listen-project

13. Chris Balma, 'As many as six billion Earth-like planets in our galaxy, according to new estimates', University of British Columbia, 16 June 2020, science.ubc.ca/news/many-six-billion-earth-planets-our-galaxy-according-new-estimates

14. Carl Sagan, via Sophie Allan, 'Happy Birthday Carl Sagan!', National Space Centre, 11 November 2020, spacecentre.co.uk/blog-post/happy-birthday-carl-sagan

—T

1. Finlay Greig, 'Space Junk: how much debris is orbiting Earth, where is it and what can be done to clean it up', iNews, 12 September 2019, inews.co.uk/news/science/space-junk-map-debris-earth-where-how-clean-up-337880

2. 'Space Debris', NASA Headquarters Library, NASA, www.nasa.gov/centers/hq/library/find/bibliographies/space_debris

3. 'Kessler Syndrome,' *Space Safety Magazine*, Michelle La Vone, www.spacesafetymagazine.com/space-debris/kessler-syndrome

4. Dan Falk, '2 large pieces of space junk nearly collided in "high risk" situation', *National Geographic*, 15 October 2020, www.nationalgeographic.com/science/2020/10/two-large-pieces-of-space-junk-have-a-high-risk-of-colliding

5. Marcia Smith, 'The Chinese ASAT Test – Nine Years On,' SpacePolicyOnline.com, 11 January 2016, spacepolicyonline.com/news/the-chinese-asat-test-nine-years-later

6. Deborah Byrd, 'Astronauts shelter as ISS boosted to avoid orbital debris', EarthSky, 23 September 2020, earthsky.org/space/astronauts-shelter-as-iss-boosted-to-avoid-orbital-debris

7. 'No one has yet been killed by re-entering space junk', *Economist*, 10 August 2019, www.economist.com/science-and-technology/2019/08/10/no-one-has-yet-been-killed-by-re-entering-space-junk

8. Kate Duffy, 'SpaceX executive says the Starship rocket system could help clean up the 760,000 pieces of space junk in orbit', 7 November 2020, Insider, www.businessinsider.com/spacex-starship-could-help-clean-up-space-junk-in-orbit-2020-11

—U

1. Robert Lawrence Kuhn, 'Confronting the Multiverse: What "Infinite Universes" Would Mean', Space.com, 23 December 2015, www.space.com/31465-is-our-universe-just-one-of-many-in-a-multiverse.html

2. Tom Fish, 'Big Bang BOMBSHELL: Are we living inside a HUGE higher dimensional black hole?', *Daily Express*, 18 June 2019, www.express.co.uk/news/science/1140059/black-hole-higher-dimensional-big-bang-theory-universe-hologram-space-news

3. Olivia Solon, 'Is our world a simulation? Why some scientists say it's more likely than not', *Guardian*, 11 October 2016, www.theguardian.com/technology/2016/oct/11/simulated-world-elon-musk-the-matrix

4. Katia Moskvitch, 'New Laser Vision Helps Telescope Probe Distant Star Cluster', Space.com, 14 May 2013, www.space.com/21148-laser-telescope-tech-star-cluster.html

5. 'History of Earth in 24-hour clock', Flowing Data, 9 October 2012, flowingdata.com/2012/10/09/history-of-earth-in-24-hour-clock

6. Jolene Creighton, 'Science Explained: How Can the Diameter of The Universe Exceed its Age', *Futurism*, 27 September 2013, futurism.com/how-can-the-diameter-of-the-universe-the-age

7. Ethan Siegel, 'Ask Ethan: How Large Is The Entire, Unobservable, Universe?', *Forbes*, 14 July 2018, www.forbes.com/sites/startswithabang/2018/07/14/ask-ethan-how-large-is-the-entire-unobservable-universe

8. Ethan Siegel, 'How Much Of The Unobservable Universe Will We Someday Be Able To See?', *Forbes*, 5 March 2019, www.forbes.com/sites/startswithabang/2019/03/05/how-much-of-the-unobservable-universe-will-we-someday-be-able-to-see

9. Oxford University, Department of Physics, https://www2.physics.ox.ac.uk/research/dark-matter-dark-energy

10. 'Big Bang', CERN, https://home.cern/science/physics/early-universe

11. 'The Early Universe', Las Cumbres Observatory, lco.global/spacebook/cosmology/early-universe

12. Lawrence Krauss, 'What Does the Future of the Universe Hold', *Smithsonian Magazine*, January 2014, www.smithsonianmag.com/science-nature/what-does-future-universe-hold-180947977

—V

1. 'Planetary Voyage', Jet Propulsion Laboratory, voyager.jpl.nasa.gov/mission/science/planetary-voyage

2. 'Images on the Golden Record', Jet Propulsion Laboratory, voyager.jpl.nasa.gov/golden-record

3. 'Sounds of Earth', Jet Propulsion Laboratory, voyager.jpl.nasa.gov/golden-record/whats-on-the-record/sounds

4. 'Music From Earth', Jet Propulsion Laboratory, voyager.jpl.nasa.gov/golden-record/whats-on-the-record/music

5. Timothy Ferris, 'How The Voyager Golden Record Was Made', *New Yorker*, 20 August 2017, www.newyorker.com/tech/annals-of-technology/voyager-golden-record-40th-anniversary-timothy-ferris

6. 'Greetings to the Universe in 55 Languages', Jet Propulsion Laboratory, voyager.jpl.nasa.gov/golden-record/whats-on-the-record/greetings

7. The Vinyl Factory, 'The story of the Voyager Golden Record', YouTube, 25 April 2018, www.youtube.com/watch?v=Mx0eNqINNvw

8. 'Oort Cloud', NASA, solarsystem.nasa.gov/solar-system/oort-cloud/in-depth/; Nadia Drake, 'Both of NASA's Voyager spacecraft are now interstellar. Where to now?', 10 December 2018, www.nationalgeographic.com/news/2017/09/voyager-40-years-nasa-interstellar-space-science

9. 'Howdy Stranger', NASA, 19 August 2002, www.nasa.gov/missions/deepspace/MI_CM_Feature_02.html

—W

1. Roxanne Oclarino, 'Defying gravity women in space', 26 April 2022, www.iso.org/contents/news/2022/04/defying-gravity-women-in-space.html

2. Louise Walsh, 'Journeys of Discovery', University of Cambridge, www.cam.ac.uk/stories/journeysofdiscovery-pulsars

3. 'NASA What is a Pulsar?', NASA Goddard, 10 February 2014, www.youtube.com/watch?v=gjLk_72V9Bw

4. Martin Durrani, 'Overlooked for the Nobel: Jocelyn Bell Burnell', *PhysicsWorld*, 30 September 2020, physicsworld.com/a/overlooked-for-the-nobel-jocelyn-bell-burnell

5. Jane J. Lee, '6 Women Scientists Who Were Snubbed Due To Sexism', *National Geographic*, 19 May 2013, www.nationalgeographic.com/news/2013/5/130519-women-scientists-overlooked-dna-history-science

6. Louise Walsh, 'Journeys of Discovery', University of Cambridge, www.cam.ac.uk/stories/journeysofdiscovery-pulsars

7. 'Nancy Roman (1925–2018)', NASA, solarsystem.nasa.gov/people/225/nancy-roman-1925-2018

8. Jamie Carter, 'The NASA Space Telescope The Trump Administration Tried To Kill Could "Find 1,400 New Planets"', *Forbes*, 26 February 2019, www.forbes.com/sites/jamiecartereurope/2019/02/26/the-nasa-space-telescope-the-trump-administration-tried-to-kill-will-find-1400-new-planets

9. Hanneke Weitering, 'NASA Renames Next-Generation Telescope after Nancy Grace Roman', *Scientific American*, 21 May 2020, www.scientificamerican.com/article/nasa-renames-next-generation-telescope-after-nancy-grace-roman

10. Leila McNiel, 'The "star-fiend" who unlocked the Universe', BBC, 12 March 2021, www.bbc.com/future/article/20210310-the-star-fiend-who-unlocked-the-universe

11. 'Vera Rubin and Dark Matter', *Cosmic Horizons Curriculum Collection*

12. Sue Nelson, 'The Harvard computers', *Nature*, 455, 36–37 (2008), doi.org/10.1038/455036a

13. Jenni Avins, '"Devise your own paths": The enduring wisdom of Vera Rubin, groundbreaking astronomer and working mother', *Quartz*, 27 December 2016, qz.com/873005/vera-rubin-quotes-wisdom-from-a-groundbreaking-astronomer-and-working-mother

14. Camila Domonoske, 'Vera Rubin, Who Confirmed Existence Of Dark Matter, Dies At 88', NPR, 26 December 2016, www.npr.org/sections/thetwo-way/2016/12/26/507022497/vera-rubin-who-confirmed-existence-of-dark-matter-dies-at-88

—X

1. Boston Dynamics, 'Do You Love Me?', YouTube, 29 December 2020, https://www.youtube.com/watch?v=fn3KWM1kuAw

2. Sam Rae and Lisa Hendry, 'What killed the dinosaurs', National History Museum, www.nhm.ac.uk/discover/dinosaur-extinction.html

3. Kasandra Brabaw, 'Avoiding "Armageddon": Asteroid Deflection Test Planned for 2022', Space.com, 18 May 2015, www.space.com/29427-asteroid-deflection-armageddon-test-aida.html

4. What If, 'What if an asteroid hit earth?', YouTube, 20 November 2018, www.youtube.com/watch?v=HFjCBstBnj8; 'What if the dinosaur-killing asteroid hit earth today?', RealLifeLore, 13 November 2019, www.youtube.com/watch?v=yeVHRjwe3F4

5. TED, 'Martin Rees: Can we prevent the end of the world?', 25 August 2014, www.youtube.com/watch?v=tMSU6k5-WXg

6. Rory Cellan-Jones, 'Stephen Hawking warns artificial intelligence could end mankind', BBC, 2 December 2014, www.bbc.com/news/technology-30290540; 'When Robots Attack: Examining Artificial Intelligence, Autonomy, and Unmanned Threats', OSAC, 2018 OSAC Annual Briefing, Research & Information Support Center, www.nmhc.org/globalassets/advocacy/isac/osac--ai-and-unmanned-threats.pdf

7. SXSW, 'Elon Musk Answers Your Questions! SXSW 2018', YouTube, 12 March 2018, www.youtube.com/watch?v=kzlUyrccbos

8. Clemency Burton-Hill, 'The superhero of artificial intelligence: can this genius keep it in check?', *Observer*, 16 February 2016, www.theguardian.com/technology/2016/feb/16/demis-hassabis-artificial-intelligence-deepmind-alphago

9. Lucy Colback, 'The impact of AI on business & society', *Financial Times*, 16 October 2020, www.ft.com/content/e082b01d-fbd4-4ea5-a0d2-05bc5ad7176c; 'Scientific advances, real world impact', Deepmind, deepmind.com/impact

—Y

1. Matt Williams, 'What is the Life Cycle Of The Sun?', 22 December 2015, Universe Today, www.universetoday.com/18847/life-of-the-sun; 'From a Dust Cloud to a Black Hole, Here's What You Need to Know about a Star's Life Cycle', Superprof, www.superprof.co.uk/blog/what-is-a-star-how-is-it-formed

2. Tim Sharp and Ailsa Harvey, 'How big is the Sun?', Space.com, 21 January 2022, www.space.com/17001-how-big-is-the-sun-size-of-the-sun.html

3. Jake Parks, 'The most extreme stars in the Universe', 23 September 2020, astronomy.com/magazine/news/2020/09/the-most-extreme-stars-in-the-universe

4. Diane K. Fisher, 'The Sun's Layers and Temperatures', January 2004, NASA, www.jpl.nasa.gov/nmp/st5/SCIENCE/sun.html; Daisy Dobrijevic, 'How Hot Is The Sun', Space.com, 21 January 2022, www.space.com/17137-how-hot-is-the-sun.html

5. Morgan McFall-Johnsen, 'The most detailed photos and videos of the Sun's surface ever captured reveal Texas-sized cells of boiling plasma', 30 January 2020, *Business Insider*, www.businessinsider.com/sun-surface-photos-video-reveal-boiling-plasma-2020-1?r=US&IR=T

6. Christopher Crockett, 'What are coronal mass ejections', EarthSky, 9 December 2020, earthsky.org/space/what-are-coronal-mass-ejections

7. 'Solar Orbiter', ESA, www.esa.int/Science_Exploration/Space_Science/Solar_Orbiter

8. Matt Williams, 'What is the Life Cycle Of The Sun?', Universe Today, 22 December 2015, www.universetoday.com/18847/life-of-the-sun/; 'The life cycle of stars', BBC, www.bbc.co.uk/bitesize/guides/zwv8xfr/revision/1

9. Matt Williams, 'Will Earth survive when the Sun becomes a red giant?', Phys.org, 10 May 2016, phys.org/news/2016-05-earth-survive-sun-red-giant.html; J. D. Meyers, 'Imagine The Universe: White Dwarf Stars', NASA, 17 November 2014, imagine.gsfc.nasa.gov/science/objects/dwarfs2.html

10. Illinois State University. '"Black dwarf supernova": Physicist calculates when the last supernova ever will happen' ScienceDaily, 12 August 2020, www.sciencedaily.com/releases/2020/08/200812113354.htm; Jean Tate, 'Black Dwarf', Universe Today, 23 September 2009, www.universetoday.com/41096/black-dwarf

—Z

1. John A. Ball, 'The Zoo Hypothesis', *Icarus*, Vol. 19, 3 (1973), pp. 347–349.

2. 'Neil deGrasse Tyson Is Worried That Humans Are Too Stupid For Aliens', *Business Insider*, YouTube, 13 July 2013, www.youtube.com/watch?v=Tt0uV5d8tss

3. Duncan H. Forgan, 'The Galactic Club or Galactic Cliques? Exploring the limits of interstellar hegemony and the Zoo Hypothesis', Cambridge University Press, 28 Nov 2016, www.cambridge.org/core/journals/international-journal-of-astrobiology/article/abs/galactic-club-or-galactic-cliques-exploring-the-limits-of-interstellar-hegemony-and-the-zoo-hypothesis/9E637E84C76BFF2FD2EDADC63CA3D1B6; 'Prime Directive', Wikipedia, Last updated: 8 November 2022, en.wikipedia.org/wiki/Prime_Directive

4. Matt Williams, 'Beyond "Fermi's Paradox" VIII: What is the Zoo Hypothesis?', Universe Today, 31 August 2020, www.universetoday.com/147573/beyond-fermis-paradox-viii-what-is-the-zoo-hypothesis/; Seth Shostak, '"Zoo hypothesis" may explain why we haven't seen any space aliens', 3 April 2019, SETI Institue, www.seti.org/zoo-hypothesis-may-explain-why-we-havent-seen-any-space-aliens; Robert Lamb, 'The Zoo Hypothesis: Are Aliens Watching Us Like Animals in a Zoo?', *How Stuff Works*, 23 November 2021, science.howstuffworks.com/space/aliens-ufos/zoo-hypothesis.htm

5. João Pedro de Magalhães, 'A direct communication proposal to test Zoo Hypothesis', ScienceDirect, November 2016, www.sciencedirect.com/science/article/pii/S0265964616300285

6. Referenced by Michael Greshko in 'Stephen Hawkings Most Provocative Moments, From Evil Aliens to Black Hole Wagers', *National Geographic*, 2 May 2018, www.nationalgeographic.co.uk/space/2018/03/stephen-hawkings-most-provocative-moments-from-evil-aliens-to-black-hole-wagers

7. Seth Shostak, 'We just beamed a signal at space aliens. Was that a bad idea?', NBC News, 20 November 2017, www.nbcnews.com/mach/science/we-just-beamed-signal-space-aliens-was-bad-idea-ncna82244

8. John Thornhill, 'Hello, Universe. Is there anyone out there?', *Financial Times*, 4 January 2022, amp.ft.com/content/55b28de3-0678-44c2-b1d9-e12fad8a4c23

9. Emma Grey Ellis, 'METI's First Message Is a Music Lesson for Aliens', *Wired*, 16 November 2017, www.wired.com/story/metis-first-message-is-a-music-lesson-for-aliens

ACKNOWLEDGEMENTS

Much to the despair of my wife, I started talking about the idea for this book about 5 minutes after my wife went into labour with our second daughter. Even though Christiana has become largely accustomed to me saying, 'I've got an idea about something,' I do agree that contractions aren't a solid foundation for a brainstorm.

As with many things in life, an idea is much easier than the application and I didn't complete the book until over 5 years later – the bulk of which took place during lockdown(s).

Without sounding too much like an acceptance speech for an award that I've neither won, nor been nominated for, there's a great deal of people who I am extremely grateful towards for their varying levels of input. Every single person who pledged for this book despite it taking me an absolute age to finish, you are a brilliant bunch and I hope the wait was worth it! Alex and Martin – thanks for taking the time to read through the advanced drafts and feed back with such detail – your unique blends of general interest, humour and not being completed terrified of quantum physics is much appreciated.

Andreas, thank you for not only indulging all my emails / WhatsApps, late night conversations and thoughts on this time consuming labour of love, but also helping to make this more beautiful than I could have imagined. I'm lucky to have such a talented friend and brother-in-law to collaborate with. I look forward to the next one.

Thanks to everyone at Unbound who has contributed along this process and helped to make this something we can look back on and be proud of. There are too many of you to thank in total, but some key people include Flo, Alex and Mark. You've all played integral roles and I'm very grateful.

To Christiana, your patience, intelligence, love and endless reserves of support will never go unappreciated or unreciprocated.

Xanthe and Polly, I made this book to make you proud and to hopefully inspire a spark of interest in space. Even if I achieve one of those things, it was all worth it. Your minds are beautiful and bold and shine as brightly as anything contained in these pages.

James Wilkins is a creative director who lives in Surrey with his wife, two daughters and dog. After spending a number of years on this project, preoccupied with the workings of the Universe, it's unlikely he'll ever write a book about space ever again.

Andreas Brooks is a British Cypriot graphic designer and visual artist currently based in Athens, Greece.

Advanced Research Projects Agency, 76

Aglietti, Guglielmo, 104

15760 Albion, 65

alien life, 6, 13, 41, 96, 100, 101, 124, 132–3

    see also intelligent life

Allen Telescope Array (ATA), 100

al-Shatir, Ibn, 24

al-Urdi, Mu'ayyad al-Din, 24

Andromeda Nebula, 72

Apollo 11, 76, 77

Apollo 14, 83

Aristarchus of Samos, 24

Armstrong, Neil, 76

artificial intelligence (AI), 125

asteroids, 12, 13, 40, 41, 87, 124, 125

astronomical units, 64

astronomy, 12–13

    astrobiology, 13, 100

    astrogeology, 13

    astrophysics, 12

    cosmology, 12–13

    extragalactic astronomy, 13

    observational astronomy, 12

    planetary astronomy, 12

    stellar astronomy, 13

Axiom Space, 55

Babylonians, 12, 58

Baggott, Jim, 93

Ball, John, 132

barycenters, 58

Basque Witch Trials, 24

Beaver, David, 82

Běhounková, Marie, 59

Bell Burnell, Jocelyn, 120

Berners-Lee, Tim, 68

Bezos, Jeff, 83

Billings, Lee, 45

biosignatures, 100

black dwarfs, 129

black holes, 6, 7, 13, 18–20, 30, 44, 45, 51, 68, 72, 77, 110, 129

Blue Origin, 83

Bohr, Niels, 92

Boston Dynamics, 124

Breakthrough Listen Initiative, 101

Breakthrough Prize in Fundamental Physics, 120

Bryson, Bill, 31

Burney, Venetia, 86

Burns, Dan, 44

Caplan, Matt, 129

carbon, 19, 129

Carroll, Sean, 92

Carter, Jimmy, 115

cave paintings, 12

Centre for the Study of Existential Risk (CSER), 125

Ceperley, Daniel, 104

Ceres, 65

CERN, 68, 69

Cetus constellation, 100

Clipper, 59

Columbus, Christopher, 133

Coma Cluster, 30

comets, 6, 13, 40, 54, 58, 64, 87, 101

Copernican revolution, 24–5

cosmic microwave background, 110

cosmological constant, 35

Cox, Brian, 68

Cretaceous period, 72

Curiosity, 13, 97

dark energy, 30, 34, 35, 110–11, 120

dark matter, 13, **30–1**, 34, 35, 69, 110–11, 121

DeepMind, 125

deGrasse Tyson, Neil, 35, 86, 87, 132

Democritus, 72

Dirac, Paul, 92

Doppler effect, 34

Drake, Frank, 40, 100, 114

Drake equation, 40

dwarf planets, 64, 65, 86

Earth, formation of, 111

Earth-like planets, 73

Edgeworth, Kenneth, 64

Egyptians, ancient, 128

Ehman, Jerry, 101

Einstein, Albert, 12, 18, 20, 34, 35, 44, 45, 92, 93

Englert, François, 69

Eridanus constellation, 100

Eris, 65

Europa, 59

European Space Agency, 50, 54, 77, 128

Event Horizon Telescope, 20

existential risks, *see* X-risks

expansion, 30, 34–5, 45, 50, 51, 72, 77, 110, 111, 121

extraterrestrial intelligence, *see* intelligent life

Fermi, Enrico, 40

Fermi paradox, 40–1, 132

Ferris, Tim, 114

Feynman, Richard, 92

Fountain, Glen, 65

Future Circular Collider (FCC), 69

Gaia space observatory, 72, 73

Galilean moons, 59

Galileo Galilei, 24, 25, 59, 72

gamma-rays, 41

Garber, Stephen J., 100

Garrett, Michael, 41, 101

general relativity, 18, 34, 45

Genzel, Reinhard, 18

geocentric model, 24

Ghez, Andrea, 18, 20

Golden Record, 114–15

'Goldilocks zone', 40

Goya, Francisco de, 129

GPS satellites, 45, 105

gravitational lensing, 19, 30, 44, 51

gravitational waves, 7, 12, 20, 44

gravity, 6, 18, 19, 20, 31, 35, 44–5, 54, 55, 59, 83, 86, 110, 114, 129

    microgravity, 54, 55

    zero gravity, 54, 55, 83

Grazier, Kevin, 58

Great Filter theory, 40–1

Greeks, ancient, 12

Hadfield, Chris, 54, 82

Harvard College Observatory, 121

Harvard–Smithsonian Center for Astrophysics, 13

Hassabis, Demis, 125

Haumea, 65

Hawking, Stephen, 19, 125, 133

Hawking radiation, 7, 19

Hayden Planetarium, 86

Heisenberg, Werner, 92

heliocentric model, 24, 25, 128

helium, 12, 58, 128, 129

Hewish, Antony, 120

Higgs, Peter, 69

Higgs boson, 7, 69

High-Z Supernova Search Team, 34

HOME: A VR Spacewalk, 83

Hossenfelder, Sabine, 59

Hubble, Edwin, 34, 50, 111, 121

Hubble-Lemaître law, 34

Hubble Space Telescope, 50–1, 77, 120

hydrogen, 13, 58, 93, 128, 129

Ingenuity, 96–7

Inouye Solar Telescope, 128

InSight, 97

intelligent life, 40, 73, 100, 101, 115, 132–3

International Academy of Astronautics, 101

International Astronomical Union (IAU), 34, 86

International Organisation of Standardisation, 120

International Space Station (ISS), 6, 44, 54–5, 82, 83, 104

Io, 59

James Webb Space Telescope (JWST), 51

Japan Aerospace Exploration Agency, 55

Jewitt, David, 65

Jupiter, 25, 58–9, 65, 87, 114, 115

Kennedy, John F., 76

Kepler, Johannes, 24, 25

Kepler, Katharina, 25

Kepler space telescope, 40

Kessler, Donald J., 104

Kessler effect, 104

Koren, Marina, 19

Kragh, Helge, 92

Kuiper, Gerard, 64, 65

Kuiper Belt, 64–5, 77, 86, 87

Kunder, Andrea, 110

Large Hadron Collider, 7, 35, **68–9**

Laser Interferometer Gravitational-Wave Observatory (LIGO), 12, 20

Leavitt, Henrietta Swan, 121

Leavitt's Law, 121

Lemaître, Georges, 34, 111

LeoLabs, 104

Leonard, Frederick C., 64

Llewellyn Smith, Chris, 68

Lowell, Percival, 86

Luu, Jane, 655

M87 galaxy, 20

McConaughey, Matthew, 45

Magalhães, João Pedro de, 132

Makemake, 65

Mariner, 65, 96

Mars, 13, 64, 77, 87, 96–7

Mars Odyssey, 97

Massimino, Mike, 55

Matessa, Mike, 133

MAVEN, 97

May, Brian, 87

Mayans, 12

Mercury, 129

Messaging to Extraterrestrial Intelligence (METI), 133

Michell, John, 18

Milky Way, 13, 18, 40, 41, 72–3, 101, 128

Milner, Yuri and Julia, 101

Mitchell, Edgar, 84

Moon, 6, 24, 25, 44, 54, 76, 77, 82, 83

Musk, Elon, 64, 125

Nancy Grace Roman Space Telescope, 120

NASA, 12, 50, 51, 54, 55, 59, **76–7**, 86, 87, 96, 97, 100, 104, 120, 124

Natarajan, Priyamvada, 19

'near Earth objects', 124

Nelson, Bill, 51

Neptune, 58, 64, 86, 114, 115

New Horizons, 65, 77, 87

Newton, Isaac, 6, 18, 24, 25, 44, 45

nitrogen, 19

Nolan, Christopher, 45

Nubian Desert, 12

nuclear fusion, 93, 128, 129

Oort, Jan, 30

Oort Cloud, 64, 115

Orion Arm, 72

overview effect, 82–3

oxygen, 19, 100, 129

Pappalardo, Robert, 59

Penrose, Roger, 18

Perseverance, 13, 77, 96, 97

Pescovitz, David, 115

phosphates, 100

photons, 44, 93

Pioneer, 65

Planck, Max, 92

planetary nebulae, 129

Pluto, 64, 65, 77, **86–7**

Ptolemy, 24, 82

pulsars, 120

quantum entanglement, 93

quantum fluctuations, 35

quantum physics, 45, **92–3**

quantum tunnelling, 129

quintessence, 35

radio astronomy, improvements in, 101

Randall, Lisa, 30

red giants, 129

Red Planet (Mars), 13, 64, 77, 87, **96–7**

red supergiants, 128, 129

redshift, 34

Rees, Martin, 125

Roman, Nancy Grace, 50, 120–1

Romans, 58

Rubin, Vera, 30, 31, 121

Sagan, Carl, 101, 114

Sagittarius A* supermassive black hole, 72

Sagittarius constellation, 72

Sagittarius dwarf galaxy, 73

Saturn, 6, 58, 114, 115

Schrödinger, Erwin, 92, 93

SETI Institute, **100–1**

Shapley, Harlow, 72

Shotwell, Gwynne, 105

Siegel, Ethan, 35, 110

singularities, 18, 19, 111

Slipher, Vesto, 34, 111

Solar Orbiter laboratory, 128

Space.com, 97, 110

space debris, 104–5

space nets, 105

space-time, 12, 18, 19, 20, 30, 34, 44, 45, 69, 132

SpaceVR, 83

SpaceX, 55, 77, 105

spaghettification, 19

spiral galaxies, 13, 30, 31, 34, 72

Sputnik 1, 76

Star Trek Prime Directive, 132

stars, formation of, 111

Stern, Alan, 87

Stott, Nicole, 83

Sumerian clay tablets, 12

Sun, 128–9

    coronal mass ejections, 128

    death, 18–19

    distance to Earth, 64

    magnetic activity, 128

    nuclear fusion, 93, 128, 129

    plasma cells, 128

    size, 128

    solar flares, 41, 128

    sunspots, 25

    temperature, 128

    wobble, 58

Supernova Cosmology Project, 34

supernovae, 19, 34, 41, 51, 110, 129

superposition, 93

supersymmetry, 31

Surrey Space Centre, 104

Sutter, Paul, 40

tachyons, 93

Thorne, Kip, 45

Thornhill, John, 133

Tianwen-1, 97

Tombaugh, Clyde, 64, 86

trash, 104–5

Universe, 110–11

    age, 110

    expansion, 30, 34–5, 45, 50, 51, 72, 77, 110, 111, 121

    composition, 110

    conditions at beginning, 68

    fluctuations in density, 110

    number of galaxies, 72

    origins, 12–13

    size, 110

Uranus, 58, 86, 114, 115

UY Scuti, 128

Vakoch, Douglas, 133

Venus, 25, 100, 129

Very Large Telescope, 97

vitamin D, 128

Voyager, 7, 59, 65, 114–15

Wakata, Koichi, 55

Wall, Mike, 97

Walsh, Louise, 120

'weak force' W and Z particles, 68

Weak Interacting Massive Particles (WIMPs), 31

weightlessness, 44, 54

Wells, H. G., 96

Wheeler, John, 19

White, Frank, 82, 83

white dwarfs, 129

women, 120–1

World Wide Web, 68

Wright, Jason T., 100

Wussler, Robert, 76

X-risks, 124–5

yellow dwarfs, 128–9

Zhurong, 97

Zoo Hypothesis, 41, 132–3

Zurbuchen, Thomas, 97, 121

Zwicky, Fritz, 30

Unbound is the world's first crowdfunding publisher, established in 2011.

We believe that wonderful things can happen when you clear a path for people who share a passion. That's why we've built a platform that brings together readers and authors to crowdfund books they believe in – and give fresh ideas that don't fit the traditional mould the chance they deserve.

This book is in your hands because readers made it possible. Everyone who pledged their support is listed below. Join them by visiting unbound.com and supporting a book today.

Billy Abbott	Natalie Bertsch	Chad Carleton	Iphigenia Demetriades	Tom Fuller
Tom Abell	Jakub & Kasia Bijak	Sam Carlier	Anne Deneen	David Futter
Drew Adams	Nigel Birch	Bill Carmichael	Sam Devito	Leonardo Gada
Terry Adams	Graham Blenkin	Daniel Carney	Miranda Dickinson	Mark Gamble
Caspar Addyman	Raphael Blessley	Jonathan Carr	Samuel Dickinson	Richard Paul Gamblin
J Adkin	BlinkAfi	Anthony Carrick	Ed Digby	Jill Gamon
Samuel Agius	Adam Boita	Nicole Carty	David Dinnage	Fiona Garden
Neil Alexander	Katia Bojo	Michael Casner	Samuel Dodson	Hollie Garwood
Ryan Alexander	Ralph Bonnett	David Lars Chamberlain	Rob Doig	Oliver Gibson
Cynthia Allan	Hayley Bourne	Kenny Chapman	Pascal Domig	Martijn Giling
Peter Allan	Bruce Bowie	Deirdre Choo	Iain Donald	Tobias Gissler
Ashley Allen	Mark Bowsher	Meagan Cihlar	Kevin Donnellon	Rupert Gledhill
Catherine Allen	Luke Bowyer	David Clark	Tom Donohue	Patrick Götz
Christopher Allen	Peter Breeden	Laura Clark	Mary Alice Dooley	Emma Grae
Sally Allen	Catherine Breslin	Andrew Clarke	Paul Douglas	James Green
Rob Alley	John Breslin	Jack Clarke	Emma Doward	Aleta Greenan
Kirk Annett	Stephanie Bretherton	Matt Clerkin	Connor Doyle	Barbara Greenstein
Ellie Arnoldi	Janice Bridger	Garrett Coakley	Susie Drinkwater	Alex Griem
Oliver Arthey	Rob Brimson	Clive Cockram	Alan Driver	Ali Griffin
Martin Ashby	Mary Bromley	Jonathan Cole	Buffie du Pon	Oliver Griffiths
Adrian Ashton	Demi Brooks	Matthew Collis	Emma Dunne	Katy Guest
Nesher Asner	Jonathan Brothers	Sam & Lonnie Corea	Pascale Duval	Qahir Gulamani
Jonathan Attfield	Jocelyn Broughton	Andrew Correia	Rob Edwards	Adam Gutierrez
Charlotte Atyeo	Joss Brown	Jo Cosgriff	Eleni & Pany	Owen Haggerty
Abraham Aucamp	Brian Browne	John Crawford	Michael Elliott	Andy Hall
Suzanne Azzopardi	Chris Browne	Tom Crawford	Stan Evans	Jake Hanrahan
Grayson Bagwell	Olly Browne	Kathryn Crossingham	Harley Faggetter	Will Hanson
David Baillie	Matt Bruce	Jack Cunningham	David Fairbrother	Paul HardwareHarry
Stefán Logi Baldursson	Sean Bryan	Steven Cushnan	Chris Farrington	Amy Harrington
Phil Ball	Erica Bullivant	Dan Dalton	Sarah Faust	Pete Harris
Marina Barcenilla	Tony, Gillian	Nick Davenport	Rebecca Fennelly	Alex Harrison
Anthony Barnett	& Hann Bunn	Eve Davis	Jane Fenton	Barry Hasler
Chris Bartlett	Danielle Burnham	Dion de Boer	Sonia Fernandes	Alexander Haswell
Melissa Bavington	Aisling Callaghan	Rachael de Moravia	Steve Fleming	Ed Hawkesworth
Oliver Bayston	Braden Camilleri	James Deacon	Jemma Foster	Nicola Haynes
Betsy Bearden	Andy Campbell	Martin Dean	Clare Fowler	Daniel Heale
Chris Beddoes	Jo Candlish	Andy Dearing	Rosie Freedman	Andrew Hearse
Clara Bergdahl	Michael Card	Joanne Deeming	Monika Fried	Paul Henry

Lucy Henzell-Thomas

Jimmy Heritage

David Hicks

Tom Hobson

Emily Hodder

Molly Horler

Zack Hoynes

Scott Huggett

Matt Huggins

Deb Ikin

Ketjow Iksnizogor

Gerald Joseph Ilagan

Dmitry Isaev

Wayne Jackson

Herbert Jacobi

Kim Jarvis & Peter Taylor

Vikash Jasani

Philip Jeffree

Helen Jeffries

Simon Jerrome

Luke Jerromes

Pamela Jezard

Gavin Johnson

Laurel Johnson

Eric Jungbluth

Jasminka Karacic

Ziaul Karim

Mike Kaye

Vicki Keeley

Jan Keeling

Margaret Kelland

Danny Kelly

Steve Kelsey

Clive Kentish

Sean Keogh

Katie Khan Wood

Dan Kieran

Angela Kilgannon

The Kinsella's

Elina Lam-Gall

Phil Lamb

John Lanyon

Holly Larsen

Edward Law

Freya Marie Lawton

Liz Le Breton

Jimmy Leach

Peter Leatherland

Tamasin Little

Nikki Livingstone-Rothwell

Abbie and Lucy Lloyd

Phil Lloyd

Paul Andrew Locock-Jones

Mark Lorch

Matt Lovell

Rob Lowe

Karl Ludvigsen

Michael Luffingham

Jose Miguel Vicente Luna

Josh M

Rob MacAndrew

Catriona Macdonald

Ethan & Hope Mackie

Ava Mandeville

Hannah Marais

Jaymes Markham-Greer

Daniel Martin

Jonathan Massey

Mark Mayall

Dawood Mayet

Chris "Biggy" Maynard

Richard Mayston

Andrew McArthur

Yvonne Carol McCombie

Tom McCoy

Maeve McCrossan

Colin McCullagh

Kerry McDermott

Ian McDonald

Jamie McHale

Andrew McMillan

Liane McNeil

Garry McQuinn

Chapin Melcher

Alessandro Meteori

Kat Michalski

John Mitchinson

James Moakes

Alex Monney

Tom Moody-Stuart

Mark Morfett

Aimee Morgan

Hannah Morrisson Atwater

Robert Morrisson Atwater

Jon Moslet

Bernard Moxham

Joerg Mueller-Kindt

James Muriel

Sheriesa Naicker

Amber Nair

Stu Nathan

Carlo Navato

John New

Al Nicholson

Duncan Edward Nicol

Alexander Nirenberg

Sheila North

Chris O'Donnell

Dr John O'Hagan

Ruth O'Leary

Mark O'Neill

John-Michael O'Sullivan

Peter Orr

Camilla Marie Pallesen

Jeffrey Paparoa-Holman

Steph Parker

Chris Parry

Vinit Patel

Stuart Paterson

Steven Pendrous

John Perkins

Kayleigh Petrie

Arron Pollard

Justin Pollard

Laura Pollard

Tracey Pollard

Steve Pont

Poopy Pants Paul

Trevor Poulsum

Tom Pounder

Stephen Press

Graham Pretty

Michael Pretty

Tony Proud

Rowland Prytherch

Judith Pullman

Lam Qadhi

Andrew Rae

Daniel Rafferty

Mike Reading

Colette Reap

Chris Richards

Catherine Roberts

Wyn Roberts

Rachael Robinson

Caroline Roddis

Payton Rodman

Saxon Rodman

Chris Rogers

Wojciech Rogoziński

Alan Ross

Joshua Ross

Merrily Ross

Jasmin Rutterford Harvey

Lisa Ryan

Terry Ryan

Brian Sage

Christoph Sander

Shrikant Sawant

Janette Schubert

Jenny Schwarz

Zac Schwarz

Joseph Scicluna

René Sellies

Nik Selman

Belynda J. Shadoan

Rajiv Shah

Ati Sharma

Peter Sherrott

John R. Shibley

Alan Sims

Margaux Sloan

Duncan Smith

Karen Smith

Paul Smith

Ben Snowden

Sophie and Zara

Dave Sox

Kerri J Spangaro

Wendy Staden

Mark Stahlmann

Michael Stalker

Emily Starling

Markus Steck

Joseph Stelling

Sabine Stoffel

Jay Stoker

Stacey Stollery

Arild Stromsvag

Ian Stuart

Ander Suarez

Jack Sullivan

Kris Sullivan

Ben Sutherland

Alastair Sweetman

Alex Tanner

Lucy Taylor

David Tennant-Eyles

George Terezakis

Andrea Thoene

Hannah Thomas

Dominic Thompson

Robbie Tingey

Keith Tisdale

To Quynh Thi Dinh

Martijn Toersche

Nicky Tomkins

Chris Tomlinson

Jo Toon

Kieran Topping

Charles Maria Tor

Mike Turner

John Tuttle

Alex Tzortzi

Isobel Ulrich

Jim Unger

Clive Upton

Rens van Bergeijk

Fabio van den Ende

Sander van Dorp

Izreal Vetruvian

(Khoa Nguyen)

Sach Vohra

Ollie Wace

Sir Harold Walker

Steve Walker

Chee Lup Wan

Lee Warner

Kelvin Washbourn

Andy Way

Liz Weldrake

Brian Whelan

Paul Whelan

Alan Whitehill

James Wigley

Christiana Wilkins

James Wilkins

John Williams

Neil Williamson

Phil Williamson

Mark Willis

Michael Willis

Ludger Wilmott

Gavin Wilson

Johanna Wilson

Keeley Wilson

Stephen Wilson

Marie Winn

Family Wolpe

Charlotte Wood

Danny Wood

Mark Wood

Peter Wood

Matt Woodroof

David Yarrow

Stephanie Young